MODERN CONSTRUCTION
EQUIPMENT AND METHODS

By Frank Harris

Modern Construction Management 3rd Edition (Blackwell)
coauthor R. McCaffer

Worked Examples in Construction Management 2nd Edition (Collins)
coauthor R. McCaffer

Ground Engineering Equipment and Methods (Granada)

Management of Construction Equipment 2nd Edition (MacMillan)
coauthor R. McCaffer

Modern Construction Equipment and Methods

FRANK HARRIS

Professor of Construction Studies
Head of the School of Construction, Engineering and Technology
The Polytechnic, Wolverhampton

Longman
Scientific &
Technical

Copublished in the United States with
John Wiley & Sons, Inc., New York

Longman Scientific & Technical,
Longman Group UK Limited,
Longman House, Burnt Mill, Harlow,
Essex CM20 2JE, England
and Associated Companies throughout the world.

Copublished in the United States with
John Wiley & Sons, Inc., 605 Third Avenue, New York, NY 10158

First published 1989
Reprinted 1991

British Library Cataloguing in Publication Data
Harris, Frank, *1944–*
    Modern construction equipment & methods.
    1. Construction equipment.
    I. Title.
    624'.028

ISBN 0-582-03196-6

Library of Congress Cataloging-in-Publication Data
Harris, Frank, 1944–
    Modern construction equipment & methods/Frank Harris.
        p. cm.
    Includes bibliographies and index.
    ISBN 0-470-21320-5
    1. Construction equipment.   2. Building.   I. Title.   II. Title:
Modern construction equipment and methods.
TH900.H318   1989   88-25077
624 – dc19   CIP

Produced by Longman Singapore Publishers (Pte) Ltd.
Printed in Singapore

CONTENTS

CONTENTS

PREFACE

During the past ten to fifteen years national economies have been subjected to rapidly increased trade flows, resulting in keener competition in the design, production and distribution of goods. The construction industry has not been immune from these effects, as demonstrated by the need for both designers and contractors to seek work in the international market place.

This growing demand for higher productivity in the economy is being reflected by an increasing mechanisation of construction work. Consequently it is essential that students in particular are aware of the applications and management of construction equipment, and also that experienced professionals have a readily accessible source of information on modern construction technology.

This book has therefore been written for students and practitioners of civil engineering, construction and building. Specifically it will find a market amongst working contractors, engineers, builders, quantity surveyors, architects, specification writers, equipment and materials manufacturers, project managers, and insurance and legal advisors.

A special effort has been made to present the principles in deciding on methods of construction, devising temporary works and selecting appropriate equipment. The text is fully augmented by diagrams, tables and performance data, which together with worked examples, should help the reader to prepare costs plans and estimates of work.

To this end the book has been deliberately structured to reflect the major categories of construction equipment and methods common to general contracting and deals in order with:

1. Excavating and materials handling plant.
2. Sand and aggregate production, concrete batching and prestressed concrete.
3. Road pavement construction with the emphasis directed towards pavement construction equipment.
4. Bridgework, concentrating on temporary works structures, falsework and formwork for deck construction.

ACKNOWLEDGEMENTS

The author is especially grateful to friends and colleagues for their helpful comments on some of the specialist topics. The book has also benefited from project studies undertaken by former students. I am especially indebted to Professor R. Seeling at Aachen Technische Hochschule, West Germany who provided the inspiration and much information for the writing of this book during my study leave.

Special thanks are extended to Janet Redman who performed the arduous task of redrafting the original diagrams and sketches. Finally I wish to thank Vera Cole, Ann Smith and Meryl Murphy for cheerfully and intelligently assuming the wearisome task of typing the manuscript.

CONVERSION FACTORS

1 mile	=	1.609 km
1 in^2	=	645.2 mm^2
1 ft^2	=	0.0929 m^2
1 yd^2	=	0.8361 m^2
1 yd^3 = 0.765 m^3	=	765 litres
1 UK gallon	=	4.546 litres
1 lbf	=	4.448 newtons (N)
1 lb	=	0.4536 kg
1 kgf	=	9.807 N
1 lbf/ft^2	=	47.88 N/m^2
1 hp	=	745.7 watts (W)
1 mph	=	1.609 km/h
1 atm (bar)	=	14.7 lbf/in^2 = 0.1 N/mm^2

The units in the text reflect the current parallel use of both customary and metric systems. Many manufacturers also still make machines in imperial sizes, e.g. excavator buckets are often denominated in cubic yards and engine power quoted as horse power units. These have been maintained with the metric equivalents provided only as an aid.

NOTES

Inevitably when using mechanical equipment safety must be considered. It is, however, beyond the scope of this book to cover all the safety aspects and associated regulations with respect to the operation and maintenance of machinery.

However, some general literature is provided in the Bibliography at the end of Section 1. More detailed guidance is usually available in the safety legislation of the country of operation.

CONVERSION FACTORS

1 mile	=	1.609 km
$1\ \text{in}^2$	=	$645.2\ \text{mm}^2$
$1\ \text{ft}^2$	=	$0.0929\ \text{m}^2$
$1\ \text{yd}^2$	=	$0.8361\ \text{m}^2$
$1\ \text{yd}^3 = 0.765\ \text{m}^3$	=	765 litres
1 UK gallon	=	4.546 litres
1 lbf	=	4.448 newtons (N)
1 lb	=	0.4536 kg
1 kgf	=	9.807 N
$1\ \text{lbf/ft}^2$	=	$47.88\ \text{N/m}^2$
1 hp	=	745.7 watts (W)
1 mph	=	1.609 km/h
1 atm (bar)	=	$14.7\ \text{lbf/in}^2 = 0.1\ \text{N/mm}^2$

The units in the text reflect the current parallel use of both customary and metric systems. Many manufacturers also still make machines in imperial sizes, e.g. excavator buckets are often denominated in cubic yards and engine power quoted as horse power units. These have been maintained with the metric equivalents provided only as an aid.

NOTES

Inevitably when using mechanical equipment safety must be considered. It is, however, beyond the scope of this book to cover all the safety aspects and associated regulations with respect to the operation and maintenance of machinery.

However, some general literature is provided in the Bibliography at the end of Section 1. More detailed guidance is usually available in the safety legislation of the country of operation.

FOREWORD

THE NEED FOR CONSTRUCTION EQUIPMENT

It is estimated that 35–45% of the cost of building work is spent on materials, and in civil engineering the corresponding value sometimes approaches 35%. Traditionally, particularly in house building, much of the material has been manhandled, but as labour costs rise in relation to the costs of using mechanical equipment, it is inevitable that increased productivity will be sought on the construction site by improved applications of machinery. Witness the increasing use of concrete pumps, conveyors and not least tower cranes on all forms of construction work. However, while there is likely to be a continuing search for improved efficiency, the construction site cannot of course be operated like a factory, because of the unique character of each project. The site itself is only a temporary feature, equipment must be installed, moved around and finally dismantled in a relatively short period of time. Therefore, unless there is a continuity of work for an item on another project, there is little advantage to be gained from investing in large numbers of machines. But as applications are extended then the possibility of transfer of units from site to site will improve and facilitate more economic utilisations. It is this aspect which demands the attention of site management when balancing labour and equipment resources, and every opportunity must be taken to improve the methods of working. If this objective is firmly upheld then mechanical equipment can be deployed: (a) on increasing production and reducing costs, (b) for replacing manual operations, particularly the heavy tasks, (c) for maintaining production where labour is scarce or too expensive to employ, and (d) for reducing materials wastage.

Whenever the strategy of mechanisation has been followed, in general there has been a reduction in the amount of labour employed without a decline in the volume of production, and a corresponding improvement in materials wastage figures.

SECTION 1 EXCAVATING AND MATERIALS HANDLING EQUIPMENT

Chapter 1 **INTRODUCTION**

EARTHMOVING MACHINES

The history of mechanical equipment, particularly for excavating duties on the construction site is relatively recent. The canals of the eighteenth century were largely excavated by manual labour and the spoil transported away in horse-drawn carts. Progress was slow as a man could dig perhaps one cubic metre of earth in an hour. The railway builders of the middle 1800s saw the introduction of the first cumbersome steam-powered face shovels, which were capable of a few tens of cubic metres of production per hour. Today machines are available in many sizes to suit a multitude of different tasks and it has become almost unthinkable to carry out excavation work on even the smallest scale without mechanised assistance. However, although the tendency is towards the use of more and specialised machinery, earthmoving equipment falls roughly into only two categories: fixed-position and moving machines. The size of project, the topography, volume of earth to be removed and many other detailed factors influence the choice of type. In general, the moving machine is used for ground levelling and bulk earthmoving whilst the static-type machine is usually operated on specific tasks. *Stationary machines* (fig. 1.1) include face shovels, backhoes, draglines and grabs. The excavator loosens the soil and loads without changing position, resulting in some loss of mobility when compared with scrapers and bucket loaders, but in compensation considerably more force may be applied at the excavation face.

Figure 1.1 Fixed-position excavating machines

FACE SHOVEL

BACKACTER

DRAGLINE/GRAB

Moving machines (fig. 1.2) include bulldozers, loaders, scrapers, graders and trenching machines. The excavated material is removed, transported and deposited in a cycle, which is a particularly useful feature when large volumes of earth need to be moved over rough terrain. For example, in both civil engineering and building work the initial construction site is often uneven and

Figure 1.2 Moving
excavating machines

BULLDOZER   LOADER   SCRAPER   GRADER   TRENCHER

requires levelling. The most suitable plant for this purpose is generally selected from the group of moving machines. Specific excavation of the foundations, basement, drain trenches, etc., of the structure to be constructed can then be performed with a fixed-position excavator.

LIFTING PLANT

Lifting equipment has been used over the centuries, simple levers for example were known to the early civilisations. Ropes and pulleys have been applied with timber beams as a form of crane for most building work up to the present day. But like earthmoving machinery, steam power led to the development of heavier lifting devices, which have been slowly refined using electric motors, diesel engines, compressed air motors, etc. Today the choice of materials handling equipment is vast and includes all forms of crane, fork-lift truck, cableways, hoists, pumps and conveyors, as demonstrated in fig. 1.3.

crawler crane   derrick

mobile crane   truck-mounted   fork–lift truck
          telescopic–jib crane

truck-mounted strut-   gantry crane
jib crane

luffing tower   horizontal tower   hoist
crane   crane

rail   conveyor

cable way   concrete pump

Figure 1.3 Materials
handling equipment

The selection of the appropriate device for a lifting application is rarely as obvious as that for earthmoving duties and many interrelated factors must be considered.

MODERN DEVELOPMENTS IN EQUIPMENT

The slow but steady improvement in construction equipment, especially earth-moving machines, and truck-mounted equipment, such as mobile cranes, relates to the introduction of the diesel engine and subsequently power-shift gears and the torque converter, whereafter the cumbersome and expensive steam engine lost its standing. More recently, the development of all-hydraulic transmissions, for example on backhoes, face shovels, bulldozers, and truck-mounted cranes, has virtually rendered redundant the manufacture of the traditional rope operated machines and not least the advances made in the structural design of large cranes now make lifts of 2000 tonnes possible. However, many of the old-type machines are still widely used, thus in the earlier chapters special attention is directed towards machine configuration in order that the earlier and later developments in construction equipment technology may be understood and compared. But before the reader is cast into details, it is first necessary to introduce a review of the power sources used with the various types of machine.

POWER SOURCES

Figure 1.4 illustrates the range of torque available at the output shaft of a steam engine, internal combustion engine, gas turbine and electric motor. The graph for each is drawn to a common scale to aid comparison. It is seen that in general a reduction in rotation speed of the output shaft (i.e. lower engine

Figure 1.4 Comparison of various forms of engine power

revolutions per minute) leads to increased torque availability. The torque being the turning moment used to drive the transmission and attachments.

The available power may be calculated from the torque multiplied by the rate of turning of the output shaft. On the customary system the unit rate of working is 33 000 foot pounds per minute and is called the horse power, written hp. The equivalent metric unit is the watt (W), and 1 hp is 746 W.

It is obvious from the diagram that the maximum power and torque are not necessarily coincident at the same rate of engine revolutions.

An engine is usually described in terms of its maximum power output called the brake horse power (bhp).

TYPES OF ENGINE

(a) *Steam* powered the early machines during the last century and represented immense improvements over animal power. Even today the high torque available from the steam engine is almost unsurpassed by other forms of power. Unfortunately, the engine and boiler are cumbersome and expensive to manufacture, as a consequence steam engines have proved unpopular in recent times.

(b) *Diesel engines* were developed during the present century and have become popular only in the last forty years or so. The available torque is very much lower than from a steam engine, but with suitable gearing, considerable improvement is possible. However, the demands upon the driver as a result of continuously having to monitor the gearing are distracting and lead to reduced production.

Nevertheless, the diesel engine has gained much favour due to its compactness, low servicing and use of convenient fuel. Furthermore, diesel engines used with hydraulic power transmission fulfil both the power and manoeuvrability demands of an excavator. Rapid development is taking place in this direction.

(c) *Electric motors* show very favourable torque characteristics, but heavy expensive electric cable is required and hence only permanent, or semi-permanent fixed-in-place machines are powered in this way, e.g. tower cranes. Electric motors react very rapidly to torque requirement.

(d) *Diesel – electric motor* – a diesel engine drives an electric motor and thus no cables are needed. This method is expensive and only seldom used, e.g. large mobile cranes and excavators.

FACTORS AFFECTING EQUIPMENT SELECTION

Selecting the correct equipment for the job ideally forms part of the construction planning process and should be chosen for a particular task only after analysis of many interrelated factors. The important considerations for both earthmoving equipment and cranage include examination of:

(a) The function to be carried out: for example the choice of equipment in many instances will be dictated by the need to cope with combinations of horizontal, vertical and mobile movement. A crane can provide both horizontal and vertical movement, but is rather poor at manoeuvring, whereas most earthmoving machinery is mobile and will transport material very efficiently over long distances, but cannot be used for lifting.

(b) The capacity of the machine: the volume of material to be moved related to the time available in the construction programme is important. But for cranes and other lifting operations, the practical weight and size of the units as dictated by the design are most important factors.

(c) The method of operation: the distance and direction of travel, speed and frequency of movement, sequence of movement, state of the ground, etc., must all be taken into account.

(d) The limitations of the method: for many earthmoving operations the choice of method may be limited by the cost of temporary work facilities, for example, haul roads may not be permitted in a particular area, the establishing and dismantling costs of a crane or conveyor system may not be economic over the length of time required for the task. These are just two simple illustrations; in practice virtually every project presents different problems.

(e) The costs of the method: the most desirable method from a feasibility point of view may involve expensive and uneconomic units, which may not be available at the desired time without the incurrence of a high cost premium.

(f) The cost comparison with other methods: it is essential that all the alternatives be considered in the plan and costed.

(g) The possible modifications to the design of the project under consideration, in an attempt to accommodate the available materials handling method: for example a single large lift in a tower block, could perhaps be reduced in size. A smaller crane might then be installed. Although this procedure may appear obvious and straightforward, because of the one-off nature of most designs and construction projects the task of analysis is seldom simple. The choice of method is rarely clear cut – a compromise is the more usual result.

| Chapter 2 | **FIXED-POSITION EXCAVATING MACHINES** |

Fixed-position machines are divided into rope operated and hydraulically operated categories. The basic structural assembly of each is similar and both have a base frame and tracks supporting the digging arm and superstructure. Although both types are usually powered by a diesel engine, the mechanical parts and operating mechanisms of each are different.

A fixed-position rope operated excavator may be converted from a basic machine to operate as any one of the following types:

(a) face shovel;
(b) backacter (backhoe);
(c) grab;
(d) dragline.

Conversion to a crane, crane and fly-jib, pile driver with leaders, pile-boring machine, earth compactor, demolition tool, rock breaker is normally a standard option.

In contrast, the hydraulically operated machines are generally manufactured for a specific application, e.g. backhoe, and are not convertible into other forms. Because of their lower maintenance costs and improved performance, increasingly hydraulics are replacing ropes and gears.

BASE FRAME ASSEMBLY

The base frame consists of a sturdy welded steel structure resting on the machine axles and supports the weight of the engine, counterweight, cab, controls, gears and other mechanical parts. Today the rope operated machine is largely confined to the dragline, face shovel and grabbing crane and as such uses a system of mechanical gears or hydraulic motors to drive the tracks and attachments. The backhoe (backacter) is probably the most popular version of the fixed-position excavator on construction sites and is almost invariably of the all-hydraulic type. Both forms are powered by a diesel engine.

GEARED DRIVE FOR WHEELS OR TRACKS

Figure 2.2 shows a schematic arrangement of the transmission from the

Figure 2.1 General arrangement of a tracked excavator

Figure 2.2 Schematic arrangement of transmission machinery for a tracked crane or excavator

engine (1), to the tracks (6), rope drum (8) and slewing ring (11). One or more of the mechanisms may be selected by engaging appropriate gears (omitted from the diagram for simplification).

The engine power to drive the tracks and operate the bucket attachment on some makes of machine is directed through a torque converter to the main drive shaft (10), via a belt or chain drive (2) and speed regulation gearing (3). The torque converter is simply a fluid coupling, consisting of a series of pump blades set facing turbine blades, all housed in fluid. As the pump blades are rotated by the engine's drive shaft, a reciprocal rotation is caused in the turbine blades. Loading of the output shaft induces 'slipping' between the two sets of blades and consequent multiplication of the available output torque, as shown in fig. 2.3. Thus a variable torque demand can be accommodated at constant engine speed. It is observed that the ratio of the output to input power is at a maximum, when the output speed is approximately 0.6 of the input speed.

The power to drive the tracks is drawn off through the gears (4), (5) and (6) (fig. 2.2) and the final connection is completed through a clutch and brake arrangement on each track. The machine is steered by disengaging one track while maintaining drive on the other.

The geared drive system is cumbersome and requires frequent maintenance.

Figure 2.3
Performance of a typical
torque converter

HYDRAULIC DRIVE FOR WHEELS OR TRACKS

Figure 2.4 shows the principles of a hydraulic transmission system. The engine (1), drives a hydraulic pump (3), via a series of speed-regulating gears (2), which in turn forces oil around a closed system to drive hydraulic displacement motors attached to each of the tracks. A special rotary coupling, centred in the slewing ring, is needed to overcome the problem that the power source is mounted on the superstructure which must be able to rotate independently of the tracks and subframe, while maintaining the closed oil circuit.

A particular feature of displacement pump and motors (fig. 2.5) is the variable torque output which can be obtained from a constant engine input speed. The pistons in the pump are activated against a movable swash plate, and the hydraulic pressure is converted to torque in the motor by pressing the pistons against a fixed inclined plate. The oil is thus pumped around the closed circuit and in turn converted to torque to drive the shaft in the motor. Clearly by altering the tilt of the plate in the pump the oil flow may be increased or reduced. Figure 2.6 shows a plot of fluid pressure against oil flow, thus as the torque requirement increases the flow of oil may be slowed down and the oil pressure increased to provide the required torque. The system is always in equilibrium, and is termed 'hydrostatic'.

Figure 2.4 Schematic
arrangement of
hydrostatic drive for the
tracks of a crane or
excavator

Figure 2.5 Hydrostatic drive using a displacement pump and motor

DRIVE SHAFT

ADJUSTABLE SWASH PLATE

ROTATION

CYLINDERS

VARIABLE DISPLACEMENT PUMP

FLUID OUTLET

FLUID INLET

FLUID INLET

FLUID OUTLET

FIXED DISPLACEMENT MOTOR

ROTATION

H.P.

PRESSURE

FLOW

Figure 2.6 Relationship of oil pressure and oil flow in a variable displacement pump

This system is increasingly being preferred by manufacturers but is not universally installed on all makes of machine, and many still prefer the alternative of geared hydraulic pumps and motors.

SUPERSTRUCTURE CONSTRUCTION

The superstructure comprises the upper part of the machine, its weight is transferred to the base frame through the large slewing ring and carries the cab, engine, digging arm, bucket and counterweight. It is necessary for the superstructure to be able to turn through 360° to both dig and dump excavated material. With mechanical transmission (fig. 2.2) slewing movement is achieved by locking in gears (9) and (12) to engage the slewing ring (11) mounted on the base frame. Alternatively, when using hydrostatic drive a separate hydraulic displacement motor engages the slewing ring (see fig. 16.2).

ATTACHMENTS AND FITTINGS

ROPE OPERATION

The digging attachment, usually a bucket, is operated and controlled by means of simple rope and drum arrangements. (A similar arrangement involving a second rope drum is also required to raise and lower the boom or arm on which the bucket is hung.) Power is supplied from a diesel engine (sometimes an electric motor on very large excavators), through a gear box, or more usually nowadays a torque converter. Band pulleys (2) transfer the power via the gear system (3) through a clutch (7) to the rope drum (8). Brakes are also needed to control the drum whenever the clutch is disengaged. The drum rope passes over a series of pulleys to the bucket. The force demanded at the excavation face is automatically regulated by the torque converter, therefore the engine may be set to run at constant speed and the need to change gear is eliminated. On the more modern units much of the mechanical gearing is replaced by hydraulic motors on each drum.

HYDRAULIC OPERATION

Rope operated machines, with the exception of draglines, grabs and a few of the very large face shovels, have largely given way to hydraulic operation. The system is similar to that used to power the tracks, but the hydraulic motor is replaced by hydraulic rams to manoeuvre the arm and bucket. The system is much simpler for the driver to operate and has resulted in neater and more compact cabs. The large levers used with rope operated machines are avoided, being replaced by finger controls.

METHODS OF EXCAVATION

FACE SHOVEL

The excavator operate best from a flat prepared surface to work above the tracks against an excavation face. It serves only to loosen and load material and is mostly used in quarrying and road cuttings. Some leverage against the wall is possible, particularly with hydraulically controlled machines, enabling some faces to be tackled directly without the need for the soil to be loosened first.

Figure 2.7 Hydraulic system to operate the arm and bucket on an excavator

Rope Operated Face Shovel

The bucket is first positioned at the toe of the wall and pulled up the face, filling as it climbs. The upper part of the machine is then slewed to the dumping position, the trap opened, the bucket emptied and finally turned to its original position, ready for the next cycle.

Figure 2.8 Rope operated face shovel

The bucket

The bucket is controlled by means of ropes and pulleys, it is usually made of welded steel with bolted-on hardened teeth to improve penetration into the soil, and is usually fixed rigidly to the arm, as in fig. 2.9(a). Adjustment is possible with some types of bucket to change the angle of dig for work in different types of soil, but this must be carried out manually when the machine is in a demobilised position (fig. 2.9b) Clay soil presents problems and causes a reduction in output due to difficulty in releasing the material from the bucket.

Figure 2.9 Bucket attachments for a face shovel

(a)

(b)

Hydraulically Operated Face Shovel

Unlike the rope operated machine, the bucket is not rigidly attached to the arm, but connected to it by means of a knuckle joint and positioned with hydraulic rams. Thus the bucket is pressed into the soil with a regulated force and consequently considerably more control over the shape of the excavation face is possible than with ropes. A further advantage is that the bucket contents are discharged from the front, rather than relying upon a flap in the base of the bucket, as is the case with an operated excavator. This is a particularly useful facility when excavating clinging cohesive soils, like clay.

Figure 2.10
Hydraulically operated
face shovel

Method of Working with the Face Shovel

Rope control. The curve is radial (fig. 2.11a) and the slope of cut is dependent upon the placing of the arm, positioning of the bucket and the power of the ripping force. *The hydraulic machine* may adopt many movements, it can work like a rope machine (curve b) or claw its way up the face by pressing and relieving as in (c). The teeth act like rippers causing the loosened material to fall into the bucket. Alternatively, the full power of the machine may be applied to force out rocks and boulders (d). The action may also be applied to strip the face in layers (e) or even perform the tasks of the loading shovel (f).

Figure 2.11
Excavation profiles using
a face shovel

ROPE — HYDRAULIC

(a)   (b)   (c)   (d)   (e)   (f)

Excavating techniques

An excavation face is tackled either using the frontal or parallel approach. In the parallel method the machine follows parallel to the face of the cutting and can obviously only load the trucks from one side. The secondary face should be opened up to a depth of about 1½ times the radius of digging arm from the centre of rotation of the superstructure. The method is similar to block cutting shown in fig. 9.8 of chapter 9.

In the frontal approach the excavator simply cuts directly into the bank, until the front of the tracks have progressed up to the toe of the face. The excavator then reverses out, manoeuvres to overlap the previous areas in readiness to repeat the procedure.

Comparisons Between Rope and Hydraulically Operated Face Shovels

Hydraulic	*Rope*
1. Swivel bucket.	Fixed bucket.
2. Short arm action with potentially greater output and greater versatility.	Long arm action, causing a reduction in output.
3. Ripping power constant along the digging curve.	Full ripping power only possible at top of reach.
4. Light arrangement.	Heavy arm needed.
5. No emptying problems.	Cohesive soil causes emptying difficulties.
6. Medium-size machine, approximately 25 m³ max. bucket size.	Large machine possible, 40–50 m³ max. bucket size.

BACKACTER (BACKHOE)

The machine is used for excavating below the level of the tracks, e.g. trenches, basements, foundations and for other excavation work in confined situations and also increasingly as a small crane for handling duties in pipe-laying, installing trench sheets etc.

Rope Operated Backhoe

The excavating efficiency and profile depends upon a combination of the weight of the bucket attachment and that of the arm to provide sufficient force to penetrate the soil. The bucket is connected to the arm by means of a knuckle joint which allows a motion similar to that of the human wrist.

Hydraulically Operated Backhoe

Both the arm and the bucket are independently controlled by means of separate hydraulic rams. Hence the shape of the excavation curve may be varied to suit the situation and allows the limits of the excavation to approach much nearer to the tracks of the machine. Additionally the variable force available with hydraulic powering gives the machine a much increased capability in harder soils.

Depths up to 8 m below the tracks may be excavated with some machines. However, the under-reaming indicated in fig. 2.13 is the design limit. Naturally this could not be achieved in practice for fear of collapsing the sides of excavation.

Methods of Working with the Backhoe

(a) When excavating large and deep pits, an operating plan should be sketched out to suit the removal of the spoil and accommodate any ramps

Figure 2.12 Rope operated backhoe

Figure 2.13 Hydraulically operated backhoe

needed for the trucks and dumpers. The material is best removed in strips working up and down the length of the excavation, to a depth well within the maximum digging reach of the machine.

(b) Small pits for column bases require careful excavation and obviously production will be reduced.

(c) Drain trenches should always be excavated working from a deep to a shallow section, especially in water-bearing ground. If a sump is located in the deep section, then water will flow away from the face of the excavation. It is usual to batter (i.e. cut) back the sides of the excavation faces to improve stability. A vertical face in most soils will quickly collapse.

(d) Lifting duties. Some National Standards permit excavators to carry out lifting duties but when a safe load indicator is not fitted, the permissible Safe Working Load (SWL) is only that stated by the manufacturer in the least stable condition. Check valves also need to be incorporated in the hydraulic circuitry in case of catastrophic failure, e.g. pipeline burst.

GRABBING CRANE

Rope Operated Grab

The grab is really a crane fitted with a grabbing bucket. The excavating efficiency is dependent upon the self-weight of the attachment and the machine is therefore limited to use in fairly loose soils.

Figure 2.14 Rope operated grab

With the rope operated machine the attachment comprises two half shells hinged at the centre, which may be of the grab type (fig. 2.15a) or a clamshell (fig. 2.15b). The grab is fitted with interlocking teeth to aid penetration of the soil for genuine excavating applications, whereas the clamshell has no teeth and is used only for stockpiling very loose materials such as sand. The whole unit is hung from a crane jib on a hoist rope, sometimes called the 'holding rope'. A separate rope is used to open and close the shells and a third rope is needed to alter the position of the boom. With three rope drums to

Figure 2.15 Grab and clamshell attachments   (a)   (b)

control the arrangement is cumbersome and slow. The grabbing crane is only selected for excavating where other methods are not feasible; for example, the grab may be used in conjunction with a long boom to provide some advantage when long reaches are required, e.g. over water. The method is useful for excavating shafts and other deep confined excavations, large diameter wells and bored piles, diaphragm walls and in spoil heaps.

Hydraulically Operated Grab

Some modern developments now allow the grab to be opened and closed hydraulically, and even rotated. Clearly such improvements increase the operating efficiency and output especially when the grab is hung from a rigid arm as the teeth can then be forced into the earth to take larger bites. However, this technique reduces the loading efficiency as the height of the grab may only be altered by lifting or raising the boom. Figure 2.16 shows a typical modern arrangement to try to overcome this difficulty.

Figure 2.16
Hydraulically operated
grab

17

Characteristics of the Grabbing Crane

Hydraulic grab	*Rope grab*
1. Capable of spot digging.	Similar to hydraulic grab.
2. Excavating depth limit: approximately 14 m.	Excavating depth limited only by length of rope.
3. Excavating force directly related to hydraulic power available.	Excavating force very much dependent upon weight of grab.
4. Very limited reach.	Long reach, dependent upon jib length.
5. Difficulties in unloading when grab suspended from long arm extension.	Limited only by boom length.

DRAGLINE

The dragline may only be used as a rope operated machine, powered through mechanical gearing or with hydraulic motors on the drums and like the grab is a crane with excavator attachments, but requiring three separate rope drums – a luffing winch, hoisting winch and drag winch.

The machine is especially suitable for excavating below the level of the tracks and may be used as an alternative to the backacter in many cases. This method of excavation presents the advantage that the bucket is cast out from a boom which itself may be long, so that excavating and dumping become possible at points wide apart. Unfortunately the same action is difficult to use in confined space. Applications include mainly excavation for drain trenches, large foundations and winning underwater deposits of gravel.

Figure 2.17 Dragline

Method of Working with the Dragline

The bucket is controlled with two separate rope drums, the drag rope A and the hoisting rope B. The driver needs considerable skill to keep the two lines in unison: a rope which is given too much slack may result in tangling, catch in the works or damage the machine, etc. The excavating cycle is represented in the stages (1) to (4) in fig. 2.18. The cycle, for example, may start, say, in the 'having just dumped' position (1). The drag rope is slack and the centre of gravity of the bucket causes the bucket to hang downwards. The hoist rope is then played out and the drag rope slack is taken up until the cutting teeth begin to bite into the soil (2). The drag rope is then wound in and the bucket

Figure 2.18 Dragline
excavating cycle

filled by pulling it through the material (3). The bucket, when full, is lifted by raising the hoist rope (4) and the material is emptied in position (1) by once again providing slack in the drag rope, and the cycle repeated. The excavating angle may be altered by changing the position of the drag chains' connections to the bucket.

Characteristics of the Dragline

1. High output is possible, up to 90% of that of the face shovel in loose soils.
2. Long reach and depths below the tracks to roughly one-third the boom length may be excavated.
3. Excavation in water is possible with the use of a perforated bucket, but with much reduced efficiency and output because the driver is operating blindly and much of the material will be slurry. Roughly 20% of normal output is likely.
4. Only light to medium soil can be excavated.
5. It is difficult to cast the bucket to the exact position required and the method is therefore only appropriate for broad sweeping excavation work.
6. The system is very difficult to operate and requires the skills of a very experienced driver.

Variable Counterbalance Backhoe

To increase the reach and digging depth of the backhoe, the Priestman company have incorporated some features of the dragline. A hydraulically controlled counterweight mounted on the superstructure is used both to control the reach of the bucket and simultaneously to counter the weight of the arm. Fine control and crowd force is provided through a hydraulic ram attached between the lower arm and bucket. The direct pull of the winch (on the dragline principle) draws the bucket through the material. An outreach of 20 m can be achieved with a 1.3 m^3 bucket.

HYDRAULICALLY
CONTROLLED
COUNTERWEIGHT

Figure 2.19 Variable
counterbalance
excavator

Micro, Mini and Midi Excavators

In recent years the trend towards small-scale sub-contracting of work has encouraged the use of minaturised machines to replace manual excavation and materials-handing methods. The smallest version falls within the micro class of backhoe (fig. 2.20a), weighs between $\frac{1}{2}$ and 1 tonne and can be towed to site. Popular duties include excavation of shallow trenchwork, footings for small buildings and foundations. The mini machine (fig. 2.20b) is slightly larger at about 3–5 tonnes, and is equipped with tracks to aid mobility around the work place. Duties are similar to a standard backhoe and other attachments such as hydraulic jack hammers can be fitted to increase versatility. The midi excavator is the next class in size and weights 6–10 tonnes and mostly competes with the wheeled backhoe/loader bucket machine.

360° SLEW

Figure 2.20(a) Micro excavator

Figure 2.20(b) Mini excavator

GENERAL NOTES ON FIXED-POSITION MACHINES

(a) The rope operated machines described consist of a similar basic unit and may be quickly converted to other forms by adding the necessary attachments; for example, by changing from face shovel to backacter. For conversion to a crane, the rope and rope drums also require changing and this may take a day's work.

(b) The most common machines are in the range 18–40 tonnes basic weight (i.e. without attachments) and support excavating buckets of 0.25–1.5 m³ capacity. The output will vary depending upon the type of soil but typically a 1 m³ backhoe should on average be capable of excavating 75–100 m³/h, always provided that the machine is in good condition.

Figure 2.21 Bucket rating

PRODUCTION INFORMATION

SOIL SWELL

The production data given in fig. 2.22 represent the output after bulking or swelling caused by disturbance when the excavator bucket is being filled. The in-place or bank volume represents the soil in the undisturbed condition. Thus

bulked volume = in-place volume × swell factor

Bulking (swell) factors:

Broken rock	1.5–2.0
Gravel	1.0–1.1
Clay	1.25–1.4
Sand	1.0–1.3
Common earth	1.1–1.3

Figure 2.22 Ideal output of the face shovel

Table 2.1(a) Tracked excavators – approximate specifications

Rope operated machines								
Engine size	(kW)	(37)	(56)	(75)	(97)	(119)	(150)	(186)
	hp	50	75	100	130	160	200	250
Machine weight	tonnes	15	22	30	50	60	75	90
Fuel consumption	(litre/h)	(9)	(14)	(18)	(23)	(27)	(36)	(45)
	gal/h	2	3	4	5	6	8	10
Backhoe								
Struck bucket capacity*	(m³)	(0.38)	(0.57)	(0.76)–(0.96)	(1.14)–(1.34)	(1.53)–(1.91)	(2.68)	(3.25)
	yd³	½	¾	1–1¼	1½–1¾	2–2½	up to 3½	up to 4¼
Max. digging depth	m	5	6.5	7.5	8.0	9.0	9.0	9.5
Dragline†								
Struck bucket capacity*	(m³)	—	(0.38)–(0.96)	(0.67)–(1.34)	(0.76)–(1.91)	(0.96)–(2.29)	(1.14)–(2.67)	(1.14)–(3.06)
	yd³	—	½–1¼	⅞–1¾	1–2½	1¼–3	1½–3½	1½–4
Length of boom	m	—	21	21	27	30	33	36
Face shovel								
Struck bucket capacity*	(m³)	(0.38)	(0.57)	(0.76)–(0.96)	(1.14)–(1.34)	(1.53)–(1.91)	(2.68)	(3.25)
	yd³	½	¾	1–1¼	1½–1¾	2–2½	up to 3½	up to 4¼
Max. cutting height	m	7.0	7.0	7.0	7.5	8.5	10.0	10.5
Grab crane†								
Struck bucket capacity*	(m³)	—	(0.57)–(2.29)	(0.76)–(2.29)	(0.76)–(3.82)	(0.76)–(3.82)	(07.6)–(3.82)	(1.34)–(5.34)
	yd³	—	¾–3	1–3	1–5	1–5	1–5	1¾–7
Length of boom	m	—	21	21	27	30	33	36
Hydraulic machines§								
Struck bucket capacity*	(m³)	(0.19)–(0.48)	(0.57)	(0.76)	(1.25)	(1.67)	(2.1)	(2.5)
	yd³	¼–⅝	¾	1	1½	2	2½	3
Engine size	(kW)	(37)	(56)	(75)	(93)	(119)	(141)	(164)
	hp	50	75	100	125	160	190	220
Machine weight	tonnes	12	15	18	22	30	36	45
Backhoe								
Max. digging depth	m	5	6	7	8	8	8½	9
Face shovel								
Max. cutting height	m	5	6	7	7½	8	8	9

* Struck bucket capacity – material flush with strike-off plane. Heaped capacity – material stands inclined above strike-off plane (see fig. 2.21). The Committee on European Construction Equipment (CECE) and the Society of Automotive Engineers (SAE) in the USA specify bucket sizes. CECE rates heaped bucket capacity on a 2:1 angle of repose while SAE stipulates 1:1. Note bucket sizes in the backhoe mode should be about 75% of face shovels for stability.

† The dragline and grab are operated like cranes and the maximum load is dependent upon the operating radius. Cranes are dealt with in chapter 14.

§ The specification is very approximate for hydraulic machines, as engine power varies widely between manufacturers with similar bucket ratings. This is complicated further by the high hydraulic pressures used by some.

Table 2.1(b) Very small hydraulic excavators

		Micro	*Mini*	*Midi*
Struck bucket capacity	(m³)	0.01–0.05	0.05–0.2	0.1–0.5
Engine size	(kW)	10	20	45
Machine weight	(tonnes)	1	3–5	6–10
Backhoe				
Max digging depth	(m)	2	3	4
Face shovel				
Max cutting height	(m)	3.5	5	6

Table 2.1(c) Very large hydraulic excavators

Struck bucket capacity	(m³)	4–8	8–12	16–30
Engine size	(kW)	250	600	1700
Machine weight	(tonnes)	80	150	500
Backhoe				
Max digging depth	(m)	9	8	12
Face shovel				
Max cutting height	(m)	12	15	20

ANGLE OF SWING

The data are also based on excavating and depositing within an angle of swing of 90°. The output volumes should be multiplied by the following factors for other angles of swing:

Angle of swing	45°	60°	75°	90°	120°	150°	180°
Factor	1.26	1.16	1.07	1.0	0.88	0.79	0.71

DEPTH OF DIG/HEIGHT OF CUT

The actual output is related to the depth of dig for backhoes and draglines, and to the height of cut for the face shovel. The data in fig. 2.22 are the production rates when the machine is operating at the optimum depth. Excavators working near the limits of the range will suffer considerably reduced output. Maximum output is achieved when the height/depth of face allows the bucket to be completely filled in a single movement from the bottom to the top of the face.

JOB CONDITIONS

The production data also assume ideal conditions and uninterrupted working. For practical purposes some reduction in output is advisable.
Example:

	Production loss (%)
Weather conditions	10
Manoeuvring	8
Mechanical breakdowns	5
Operator efficiency	7
Waiting for other plant	10
	40

23

Problem
Determine the hourly production of a $\frac{3}{4}$ yd^3 (0.57 m^3) capacity backhoe, excavating a foundation 3 m deep in common earth.

Solution

Ideal output for the equivalent face shovel in fig. 2.22 for easy dig = 100 m^3/h.
Backhoe proportion of face shovel production = 0.80

Assume angle of swing is 90° therefore

Swing factor = 1.00.

Assume depth of dig allows optimum production, therefore

Adjusted production rate = 100 × 1.0 × 0.80 = 80 m^3/h.

Other factors affecting output:
Job conditions, losses caused by:

	%
Weather, say	10
Manoeuving	8
Breakdowns	5
Operator skill	7
Waiting	10
	40

Therefore

actual production of bulked material = 80 × 0.6 = 48 m^3/h

In-place production. The soil is disturbed and enters the bucket in a bulked condition during the digging process. Common earth has a swell factor of about 1.1. Therefore

$$\text{In-place production rate} = 48 \times \frac{1}{1.1} = 44 \text{ m}^3/\text{h}$$

Chapter 3   METHODS OF TRANSPORTING MATERIALS

The choice of system in transporting material from the loading point depends on many factors including:

(a) site conditions;
(b) volume of material to be moved;
(c) type of material;
(d) time available.

The choice is further complicated by considerations such as:

Will transport be over metalled roads or not?
Must the roads be kept clean?
Management impositions, e.g. bonus targets, overtime, etc.

In terms of moving excavated earth, a rough guide in selecting an appropriate system is given in fig. 3.1. The first six methods are discussed in other sections of the book, and this chapter will concentrate mostly on the means available for transport of materials over distances of several kilometres.

Figure 3.1 Transport systems for earthmoving

THE LORRY (TRUCK OR WAGON)

The wide range of trucks shown in fig. 3.2 is now available to suit almost any type of work. For transporting building materials on public highways, conventional trucks such as types (1), (2), (5) or (6) are the most appropriate and a considerable choice of designs is available to suit the goods to be carried. Such lorries distribute the load more evenly on to the axles and are designed to limit the maximum axle load within legal requirements, which currently also stipulate the maximum vehicle laden weight and body length. Wagons used for transporting excavated earth are usually of the rear dump type (3), (4) and (7) and ideally should have four-wheel drive capability to overcome both difficult ground conditions and for travelling at relatively high speeds on smooth roads. The more popular trucks tend to be supported on two axles to improve manoeuvrability and reduce the turning circle but the load on the rear axle when laden is high compared to three-axle trucks. However, modern large volume tyres, twin mounted are capable of carrying these heavy loads and the vehicle is versatile in all but very soft going ground conditions. Three-axle wagons need only be used where it is necessary to limit the bearing load such as on clays and fine soils. These trucks are more expensive than the two-axle version due to the difficulty of accommodating the large diameter wheels along a fairly short body length.

Trucks are described in terms of the total number of wheels and drive wheels. Thus a 4 × 4 truck has four wheels and four-wheel drive. An 8 × 4 truck has eight wheels, of which only four provide drive, the other four are free wheels. Sometimes wheels are mounted in pairs on the axle, to carry

	GROSS VEHICLE HP (kW)	CAPACITY (m³)	NET WEIGHT EMPTY (tonnes)	PAYLOAD (tonnes)
1	(45-261) 60-350	—	up to 10	up to 16.26 *
2	(150-261) 200-350	—	up to 20	up to 24.38 *
3	(112-1194) 150-1600	6-70	6-120	up to 150
4	(150-2235) 200-3000	6-120	10-250	up to 300
5	(75-298) 100-400	—	4-10	up to 16.26 *
6	(150-298) 200-400	—	up to 20	up to 38 *
7	(75-250) 100-325	approx. 3-20	up to 30	up to 35

*Gross laden weight only, all others are carrying load.
GLW is generally about 38 tonnes depending upon the country of operation—8 tonnes or so is usually the permitted axle load.

Figure 3.2 Transport using lorries and trucks

heavy loads. Thus although a 4 × 4 truck may perhaps have four wheels on the rear axle making a total of six tyres, the vehicle designation remains as 4 × 4. The modern rear dump truck is used for a variety of purposes from general site use up to the 300 tonnes versions in quarrying operations. The latter are frequently loaded by loader shovel, which tends to be more violent than other loading machines and requires the dump truck to be very robust.

TRANSPORTING DISTANCES

Earthmoving trucks are mainly used to transport material over distances between 1 and 10 km but longer journeys are possible. Wagons are usually loaded by an excavator, but sometimes, in exceptional cases, by continuous equipment. After completion of loading there is a short delay while positioning the next truck for loading. Obviously the waiting time for the excavator is reduced as the wagon size increases. Conversely the smaller the wagon, the shorter the loading period and the corresponding waiting time for the truck, but the optimum wagon size is very difficult to determine and depends very much on site conditions. Many workers have attempted to represent the loading/discharge cycle in a mathematical form or simulation model, but as yet the information required for input is too diffuse to enable the users to rely with much confidence on the output information. In practice, the optimum wagon size seems to be between three to ten times the size of the loading buckets, five being a good average ratio, but the final choice of truck size very much depends upon building up experience on a variety of differing types of contract, ground conditions, loading plant, etc.

HAUL ROADS

For economic transporting, haul roads should ideally be limited to a maximum gradient of 15%, preferably without bends and curves, have a hard surface and be continuously maintained in this condition by grading. The cost of a grader is easily recovered in the lower maintenance and repair costs of the trucks. When these savings are combined with the lower tyre wear and increased output resulting from the high speeds attainable, the grader proves a vital machine in the earthmoving operation.

THE DECISION TO USE LARGE DUMP TRUCKS

It is uneconomic to use very large dump trucks unless large volumes of earth are moved, but they can offer many advantages under certain conditions. For example, a reduced number of trucks means that fewer drivers are required, with a corresponding reduction in labour costs. The capacity of modern trucks now matches that of the largest loading machines, leading to less waiting time and thus a reduced operating cost. Large wagons are very robust and the exhaust system can also be arranged to provide heating and thereby prevent the soil clinging to the body. The is a particularly useful facility for carrying and emptying difficult materials.

Large rock fragments can be carried so obviating the expense of breaking the strata into small sizes.

Finally, some large trucks are now available in an articulated form which

27

affords up to a 25% saving in manoeuvring time and furthermore helps when roads with sharp bends and obstruction have to be used. They are also longer and narrower than the rigid type, so providing savings in haul road widths. Being lighter, the power-to weight ratio is better as well.

EXAMPLE OF TRUCK SELECTION

Trucks are loaded by a 1 m^3 capacity tracked loader machine at the rate of 30 m^3/h bulked material. The truck transports the material to a tip 3 km away. Select the size and number of trucks required.

Select truck size, say, loader bucket capacity × 5 = 5 m^3 capacity truck.

	Minutes
Load time = $\frac{60}{30}$ × 5	10
Cycle time to tip and back (average speed, 20 km/h)	18
Dumping, turning and accelerating	3
	31

Truck 1	10	18	3			
Truck 2		10	18	3		
Truck 3			10	18	3	
Truck 4				10	18	3

1 min waiting
for excavator

Thus three 5 m^3 capacity trucks should be adopted.

Note: The analysis in practice is more complex, and an estimate of the quality of haul road, down time, etc., is necessary before the information assumed above can be obtained. In order to include the many variables which may affect the cycle time, Monte-Carlo simulation techniques have been developed. Models are available from the major equipment manufacturers, and operators concerned with haulage at long-term sites have also developed models to suit the particular situations. The principles, however, are fairly well established as demonstrated in the following example.

EXAMPLE OF TRUCK SELECTION BY SIMULATION

Trucks arrive at an excavator from distribution points on a large earthmoving project. The arrival time intervals of the trucks are observed and yield the following results.

Arrival time interval (min)	*Frequency (%)*
2	10
3	15
4	30
5	25
6	20

The times taken to load the trucks, which are either 6 or 12 m³ capacity, are fairly constant at 3 and 5 min, respectively, and both types are equally represented in the fleet.

If the excavator loads each of the trucks immediately it arrives, in the order that it arrives, calculate the total time that the excavator and trucks will be waiting in any one period of two hours selected at random.

Solution

The trucks arrive at the excavator at varying times in an indistinguishable pattern. However, the percentage distributions of the arrival times are known, therefore it is possible to simulate the arrival time interval of a truck by drawing a random number lying in the range represented by the observed distribution. Thus:

Arrival time interval (min)	Frequency (%)	Random number
2	10	00–09
3	15	10–24
4	30	25–54
5	25	55–79
6	20	80–90

Table 3.1 simulates the arrival of 50 trucks one after the other. The random numbers are selected from table 3.2 and the type of truck arriving is determined by tossing a coin. Truck type A represents a 6 m³ load. Truck type B represents a 12 m³ load.

Table 3.1 Simulated arrival times

Random number	Arrival time (min)	Type of truck	Clock time of arrival (min) (6 m³)	(12 m³)
89	6	A	6	
29	4	B		10
73	5	A	15	
50	4	A	19	
47	4	B		23
51	4	A	27	
32	4	B		31
58	5	A	36	
07	2	A	38	
49	4	A	42	
14	3	A	45	
02	2	B		47
26	4	B		51
97	6	B		57
83	6	A	63	
47	4	B		67
51	4	A	71	
55	5	B		76
92	6	B		82
07	2	B		84
45	4	A	88	
85	6	A	94	

Table 3.1 Simulated arrival times (*contd*)

Random number	Arrival time (min)	Type of truck	Clock time of arrival (min) (6 m³)	(12 m³)
76	5	A	99	
15	3	A	102	
78	5	B		107
68	5	A	112	
83	6	B		118
33	4	A	122	
18	3	B		125
04	2	A	127	
11	3	B		130
98	6	A	136	
11	3	B		139
18	3	B		142
34	4	B		146
84	6	B		152
47	4	B		156
18	3	B		159
46	4	A	163	
71	5	A	168	
53	4	B		172
88	6	B		178
78	5	A	183	
30	4	A	187	
71	5	A	192	
35	4	B		196
96	6	A	202	
78	5	B		207
57	5	A	212	

Table 3.2 Random numbers

29	97	38	28	97	54	95	94	54	79	93	88	1	82	40	62	93	78	8	88	64	58	31	6	45
78	26	21	3	10	14	30	18	18	84	4	11	62	16	70	2	16	7	31	97	79	56	98	79	39
75	49	97	87	79	32	66	57	89	56	81	70	53	83	21	25	26	56	55	56	34	59	74	21	76
94	25	80	13	50	67	95	11	78	42	25	91	82	74	30	9	21	90	26	44	23	81	74	51	76
54	40	59	35	35	47	55	55	73	50	85	38	61	49	8	99	14	2	17	94	61	95	25	85	66
35	76	40	98	88	45	58	65	12	18	7	43	36	28	10	7	84	16	0	49	32	19	66	87	12
37	87	64	2	11	8	2	20	20	6	12	16	95	60	69	81	76	75	88	69	95	13	76	70	19
11	60	36	32	48	57	96	71	43	65	62	37	78	51	87	3	64	45	33	47	17	3	29	15	33
61	57	28	33	96	48	50	86	85	83	1	53	45	70	19	24	48	10	90	99	68	89	58	52	94
84	10	80	38	90	52	99	85	52	49	66	63	69	12	32	37	32	28	96	59	78	62	25	95	84
44	14	10	13	1	66	69	55	12	1	20	94	22	93	90	16	82	64	28	47	20	71	67	21	93
92	3	96	16	65	42	68	71	56	75	17	72	63	80	81	3	42	50	28	92	45	44	13	45	21
13	55	9	80	31	51	35	96	32	71	20	22	80	42	18	57	5	5	20	1	80	8	15	42	15
38	43	40	25	46	88	84	87	75	2	39	99	3	76	76	62	88	97	89	7	97	15	70	27	27
39	9	51	67	63	4	76	76	86	10	50	28	97	9	87	65	13	33	28	96	12	27	31	88	48
89	29	73	50	47	51	32	58	7	49	14	2	26	97	83	47	51	55	92	7	45	85	76	15	78
68	83	33	18	4	11	98	11	18	34	84	47	18	46	71	53	88	78	30	71	35	96	78	57	19
42	44	69	11	31	49	98	33	76	12	81	30	83	3	87	5	88	21	78	89	94	32	37	83	74
48	73	61	15	45	74	41	27	26	48	22	46	86	27	12	21	28	38	56	48	86	18	39	39	19
89	99	62	81	98	40	33	62	77	60	85	37	15	69	76	71	38	27	32	40	53	36	4	51	54

Start →

Table 3.2   Random numbers (contd)

83	15	51	17	86	77	66	84	50	84	44	96	92	78	37	24	49	35	54	43	78	50	40	32	56
17	25	7	90	90	70	48	70	69	23	8	85	8	95	84	53	85	55	11	93	41	6	66	72	39
23	76	93	41	27	30	77	61	72	74	81	13	73	21	99	1	47	52	44	19	51	25	29	43	54
76	57	8	22	23	26	22	70	63	70	6	6	59	75	92	86	60	50	87	81	36	80	83	43	17
96	79	7	87	51	5	17	61	43	13	64	77	45	7	55	68	20	0	17	23	64	83	61	76	37
68	23	26	10	82	97	77	2	89	33	70	46	23	45	83	99	55	95	4	41	89	33	49	89	86
19	93	65	67	40	81	96	44	68	47	78	3	18	58	1	48	19	2	34	49	99	56	54	71	65
34	76	58	86	21	86	84	70	40	23	89	26	42	62	69	10	63	32	80	30	18	12	75	34	81
26	45	91	80	51	26	64	71	6	49	96	57	56	49	81	91	77	95	44	51	61	34	89	73	78
98	26	56	88	66	51	69	71	48	14	72	40	57	32	23	54	36	66	29	10	99	4	41	86	60
53	60	78	66	81	67	45	56	64	78	19	79	10	2	55	61	32	20	20	49	11	89	54	64	96
0	99	16	71	84	95	51	9	72	70	81	23	33	89	62	20	78	77	64	37	52	39	88	16	92
24	43	49	79	28	16	10	94	49	35	95	98	59	27	70	55	34	62	91	88	56	71	79	75	31
30	14	32	60	8	33	73	62	89	63	65	18	15	22	39	29	89	8	58	11	83	66	4	15	74
39	58	83	63	94	73	84	48	95	17	79	74	78	39	10	38	35	75	74	70	69	54	82	75	50

A check on the distribution of arrival times simulated produces a reasonable fit, as shown below.

Arrival time interval (min)	Observed frequency (%)	Simulated number of arrivals	Simulated frequency (%)
2	10	4	8
3	15	8	16
4	30	17	34
5	25	11	22
6	20	10	20

It is now possible to simulate the whole loading process, since if the type of truck is known its loading time can also be determined. Table 3.3 shows the situation at the excavator over a two-hour period.

Table 3.3   Simulation situation at the excavator

Arrival type truck	Clock time of arrival (min)	Starts loading (min)	Loading time (min)	Completes loading (min)	Waiting time of excavator (min)	Waiting time of truck (min)
A	06	06	3	09	—	—
B	10	10	5	15	1	—
A	15	15	3	18	—	—
A	19	19	3	22	1	—
B	23	23	5	28	1	—
A	27	28	3	31	—	1
B	31	31	5	36	—	—
A	36	36	3	39	—	—
A	38	39	3	42	—	1
A	42	42	3	45	—	—
A	45	45	3	48	—	—
B	47	48	5	53	—	1
B	51	53	5	58	—	2
B	57	58	5	63	—	1

Table 3.3 Simulation situation at the excavator (contd)

Arrival type truck	Clock time of arrival (min)	Starts loading (min)	Loading time (min)	Completes loading (min)	Waiting time of excavator (min)	Waiting time of truck (min)
A	63	63	3	66	—	—
B	67	67	5	72	1	—
A	71	72	3	75	—	1
B	76	76	5	81	1	—
B	82	82	5	87	1	—
B	84	87	5	92	—	3
A	88	92	3	95	—	4
A	94	95	3	98	—	1
A	99	99	3	102	1	—
A	102	102	3	105	—	—
B	107	107	5	112	2	—
A	112	112	3	115	—	—
B	118	118	5	123	3	—
A	122	123	3	126	—	1
				Total	12 min	16 min

Waiting time of excavator = 12 min in 2 hours
Waiting time of trucks   = 16 min in 2 hours

The simulation should now be repeated several times, so that the results can be expressed with more confidence by the use of statistics. With the model established, and mounted on a computer, the user is in a position to experiment by changing the parameters, e.g. truck and excavator sizes, inter-arrival times to represent changes in the haul road, conditions etc.

Chapter 4   MOVING EXCAVATING MACHINES – THE BULLDOZER

The bulldozer shown in fig. 4.2 is a very versatile machine and is used frequently for:

Stripping top soil
Clearing vegetation   }   fig. 4.1(a)
Shallow excavating   fig. 4.1(b)
Pushing scrapers   see chapter 5
Maintaining haul roads   compare with the grader, fig. 6.4
Opening up pilot roads   fig. 4.1(c)
Spreading and grading   fig. 4.1(d)
Ripping   fig. 4.11

Figure 4.1(a) Stripping top soil and vegetation

Figure 4.1(b) Shallow excavation and stockpiling

Figure 4.1(c) Opening up pilot roads

Figure 4.1(d) Spreading and grading

Figure 4.2 Tracked dozer

The dozer is normally operated with a straight blade fixed perpendicular to the line of movement to push the material forward or angled to move the soil to the side.

STRUCTURAL ARRANGEMENT

The machine is assembled in two quite separate sections comprising, A, a base frame or robust welded construction to which are attached the mountings for the dozer blade, the drive sprockets and rollers for the tracks and supports for the upper body, and, B, the superstructure which carries the engine, transmission, hydraulics, cab and controls and is mounted on the base at three points 1, 2, 3 (fig. 4.3). A heavy duty equaliser bar at point 3 separates the tracks and allows independent movement for better distribution of the machine's weight and improved traction when working on uneven ground.

Figure 4.3 Structural form of tracked dozer

TRANSMISSION

Bulldozers are usually powered by a diesel engine with an output of about 60 hp (45 kW) for small machines, up to 700 hp (530 kW) for the heavy duty dozer used for ripping operations. Sometimes a mechanical clutch is fitted between the engine and gears (fig. 4.4) which even today provides the following advantages over other forms of transmission:

(a) Cost: the simple arrangement tends to result in a cheaper machine.

(b) The engine may be throttled down in order to produce more torque. This is a very useful facility when concentrating on forcing out large stones and rocks, but a manual gear change demands a high concentration from the driver in selecting the correct gear to prevent an engine stall during varying load conditions. Many drivers tend to reduce the efficiency by taking a lower gear than necessary. Today most modern bulldozers employ a torque converter and power-shift gears as shown in fig. 4.4. The primary clutch is eliminated so dispensing with the need to disengage the engine during changing gear, as all the gear wheels are in connection. The gear requirement is altered by engaging the relevant secondary clutch available on each gear ratio. The torque converter provides these advantages compared with the mechanical clutch and gear box, e.g.

  (i) shock loading when changing gears is eliminated;

  (ii) full torque can be applied almost immediately as the engine works at constant power output. The engine speed will vary only marginally by about plus or minus 50 rev/min.

(c) The torque converter automatically adjusts the torque requirements in the selected gear to suit the working conditions, namely:

high torque requirement = low output speed
low torque requirement = high output speed.

(d) There is no primary clutch, so the driver can concentrate solely on the work. Such a machine may be effectively operated by a less skilled driver than is needed for a clutch and gearbox machine.

ALTERNATIVE
TORQUE CONVERTER AND POWER-SHIFT GEARS
FOR A SMALL TRACKED DOZER, SHOWING FIRST
GEAR ONLY.

Figure 4.4 Clutch and gears for a small tracked dozer

Figure 4.5(a) Dozer pivoted around one track

Figure 4.5(b) Dozer turned by slipping one track

Figure 4.5(c) Dozer pivoted at centre of machine

Some very modern machines are now manufactured with all-hydraulic transmission and control of the attachments. This system uses displacement pumps and motors as described earlier in chapter 2. The machines have not as yet, however, proved robust enough to cope with heavy duty work and further development is taking place.

STEERING

Positioning is achieved by disengaging the track clutches, one for each track.

(a) For sharp turning, one track is used as the pivot point (fig. 4.5a). The clutch to this track is disengaged and the brake applied, full drive is maintained on the other track.

(b) A broader turning circle is obtained by not completely disengaging the clutch on the pivot track as in fig. 4.5(b). Unfortunately, both these methods tend to cause much clutch and track wear.

(c) A very small turning circle may be obtained on machines with the facility for allowing the tracks to rotate in opposite directions (fig. 4.5c). Such a feature is not required for normal operation, but the extra cost is justified for work in confined areas, such as tunnelling.

THE WHEELED DOZER

The basic machine is similar in construction to the wheeled loader with the bucket attachment replaced by a bulldozer blade. Four-wheeled drive is required to give sufficient traction to perform the tasks of a bulldozer and a correspondingly larger engine than that needed for the tracked dozer is necessary. The increased self-weight in turn raises the bearing pressure at the point of contact between tyres and ground, causing compaction of the soil which may be undesirable in view of the fact that to be efficient a dozer requires loose material to push. Furthermore, the close centres of the two axles may cause stability problems on uneven ground. Wheeled dozers are seldom used today for these reasons.

Figure 4.6 Wheeled dozer

METHODS OF WORKING WITH THE BULLDOZER

THE BLADE

The dozer is fitted with a stiff welded steel blade and pushes material forward, transferring force from the tracks to the blade by means of two strong arm

Figure 4.7 Angled dozer blade

connections to the base frame (see fig. 4.2). The blade is controlled with two sets of double hydraulic cylinders. One pair connected to the arm attachments (1) act to tilt the blade and in doing so control the depth of dig. The other pair of cylinders are attached to the upper frame (2) and serve to lift and lower the blade for digging effect.

The blade may be angled (fig. 4.7) to deposit the material at one side along the push line which is a very useful facility when opening up pilot roads or backfilling trenches.

Straight Blade

Most blades are curved but the section perpendicular to line of push (fig. 4.9a) is straight. The curvature causes the material to roll forwards (fig. 4.8). The driver controls the depth of cut during the pushing action by feel. When the rear of the dozer is felt to rise, the blade is starting to dig in, the angle and depth of the blade are then 'tuned' to meet the resistance encountered. The volume of material that can be moved forward by the blade is dependent upon the blade size and shape. The maximum length of push should not exceed 100 m, as the machine is uneconomical for earthmoving over greater distances.

Figure 4.8 Flow of material in front of a dozer blade

U-Blade

The blade in cross section has a much deeper curvature approaching almost a U-shape (fig. 4.9b), in addition the outer edges are angled slightly inwards. This type of blade will carry much larger volumes of material, but is limited to light, flowing-type soils.

Angled Blade

The blade in plan view is angled and casts the material to one side. The arrangement is used mainly for backfilling along a trench and for opening up pilot roads in hilly terrain and not as a general purpose earthmover as with the straight-ahead blade.

Figure 4.9 Bulldozer
blades

(a) STRAIGHT BLADE

(b) U BLADE

(c) ANGLED BLADE

(d) PUSH BLADE

Push Blade

Push loading a scraper is one of the main functions of the bulldozer particularly for the larger machines (300 hp (223 kW) upwards). Modern motorised scrapers are only self-loading in very light soils and often must be assisted to load by a bulldozer. The blade (fig. 4.9d), usually fitted with shock absorbers, effective at travelling speeds up to about 5 km/h, is of much stronger construction than the normal blade and is stiffened with a push plate.

Table 4.1 gives approximate data on the size of blade generally adopted with a given size of machine.

Table 4.1 Bulldozer blades

Engine size (hp)	700	450	400	335	300	200	100–150	60–70
(kW)	520	340	298	250	224	149	75–112	45–52
Machine weight (tonnes)	85	50	45	35	25	20	10–14	5–8
Blade length (m)	5.5	5.0	4.5*	4.0	4.5	3.5	3.5	3.0
Blade height (m)	1.8*	1.8*	1.8*	1.5	1.5	1.2	1.0	0.8
U-blade (m³)	20	18	14	13	9	7	4	2

* Not suitable for ground surface trimming because of size causing problems of control.

PRODUCTION DATA

The production information shown in fig. 4.10 is ideal output assuming favourable conditions. The values should be adjusted to suit the circumstances of the construction site as explained in chapter 2.

RIPPING ROCK

RIPPER

The principle of ripping hard soil and rock was known in the 1930s but only with the development of powerful bulldozers and robust tracks has the method become a practical proposition.

Figure 4.10 Ideal
output of the bulldozer

Production data

Assumptions

(i)	Dozing speed	3 km/h
(ii)	Return speed	6 km/h
(iii)	Manoeuvring	0.15 min
(iv)	Uninterrupted production	
(v)	Volumes are bulked, in-place quantities should be reduced by an appropriate swell factor	

NOTE

For operation in other soil multiply
production output by the following

Loose soil	–	1.00
Firm soil	–	0.85
Broken rock	–	0.65

STRAIGHT BLADE
CAPACITY

8 yd³	6.11 m³
6 yd³	4.58 m³
4 yd³	3.06 m³
3 yd³	2.29 m³
2 yd³	1.53 m³
1·5 yd³	1.14 m³

Until the development of the modern ripper, soil or rock which could not be excavated with traditional excavating equipment had to be blasted loose and into sizes which were easily disposable, but today large quantities of rock may be removed economically by ripping methods.

SELECTION OF RIPPING EQUIPMENT

The ripping effect requires (a) penetration into the soil of a strong steel ripping tool, (b) a tractor with sufficient horse power to enable the ripping tool to advance the point through the material, (c) a heavy tractor to generate sufficient traction, (d) a strong robust tractor to take the strain. The ripping tool used is normally a radial are type or parallel linkage type.

Figure 4.11 Ripping
methods

(a)

(b)

39

Radial Arc Type

This is of simple construction (Fig. 4.11a) and requires little servicing, but because the ripping tool approaches the digging action along an arc, there is a tendency for the rear of the tractor to lift and thus diminish the available traction.

Parallel Linkage Type

The ripping tool enters the ground upright, allowing the point to immediately dig in and thus pulls the tractor to the ground to increase traction. This method (Fig. 4.11b) is probably the better of the two systems, especially in tough materials. The latest versions can now be equipped with hydraulic vibrating mechanisms for improved penetration.

Shape of Shank

The shank is made from high tensile steel usually fitted with a replaceable point. For hard compact material and laminated rock the straight shank (fig. 4.12a) gives good performance. The curved shank (fig. 4.12b) is more suited to ripping fractured rocks.

(a)   (b)

Figure 4.12 Ripping tool shanks

RIPPING TECHNIQUES

(a) The minimum tractor power required is 150 hp (112 kW), but for very tough work 400 hp (300 kW) or more may be needed.

(b) First gear should always be used, travelling at 1–2 km/h to obtain maximum torque from the torque converter.

(c) When conditions permit, it is advisable to rip with three teeth to reduce speeds to the least possible and minimise wear and tear not only to the ripping tool but also to the machine itself. The number of teeth is best chosen by trial and error but when working with material which tends to break out in large slabs, generally only one point is necessary.

(d) Modern teeth can now rip to depths approaching one metre, the limiting depth depending very much on the type of ground. Usually it is desirable to rip as deep as possible.

(e) The distance between passes is again largely governed by the type of ground. As a guide, the following may be suitable:

hard material   1–1.5 m passes
light fractured material   2–2.5 m passes.

(f) The material should be removed either by scraper or loader.

(g) Output estimation is very difficult to assess and the only reliable method is to carry out site observation. For tendering purposes the figures shown in fig. 4.13 which are based on records collected by Pohle at Aachen T.H., Germany, are very approximate guides.

(h) Ripping methods may offer up to 80% saving compared to the cost of using explosives.

Rocks with Characteristics Favourable to Ripping Methods

(a) Stratified rocks and soils.
(b) Rocks with fractures, faults and planes of weakness.
(c) Large-grained rock of brittle texture.

Figure 4.13
Theoretical output from
a rippable rock

(d) Most shales, slates and mudstones.
(e) Soft rock.

Rocks with Unfavourable Characteristics

(a) Non-crystalline rocks.
(b) Igneous rocks in general, but specifically, those with massive and uniform structures.
(c) Cohesive soils.
(d) Rocks without planes of weakness, and soils without bedding planes.

It is now possible to obtain an indication of the rippability of the rock by seismological methods (see reference 8, Bibliography: Section 1).

OTHER ATTACHMENTS

TOWING WINCH

The larger machines often have a winch attachment at the rear, which serves for stump pulling, moving sheet piling and general skidding jobs.

SIDE BOOM

Side booms have been developed mainly for use in the handling and laying of large diameter pipes. Boom lengths up to 6 m are available and loads up to 30 tonnes may be lifted with very large machines (400 hp (248 kW)). The modern machines have a hydraulically variable positionable counterweight, with independent boom and lift ropes.

Figure 4.14 Side boom attachment for a tracked dozer

MARSH TRACKS

Bulldozers especially developed for operating on marsh ground have very wide tracks. Thus the contact area of the track is increased and bearing pressure is reduced:

bearing pressure marsh tracks   approx. 0.03 N/mm^2
bearing pressure normal tracks   approx. 0.05 N/mm^2
bearing pressure under human foot approx. 0.02 N/mm^2.

Chapter 5   MOVING EXCAVATING MACHINES – THE SCRAPER

INTRODUCTION

The bulldozer can be used effectively for moving earth over short distances up to 100 m. However, many projects, particularly road construction, necessitate a combined load, haul and discharge system at least up to a distance of 3 km. The situation calls for a robust excavator, capable of travelling over rough terrain to eliminate the need for the use of trucks and wagons on public roads. The scraper has been developed specifically to cater for this medium-distance haul. Essentially the earth is cut and loaded directly into the scraper box (or bowl), transported to the discharge area and finally spread in layers. The whole process takes place in a continuous cycle. The type of machine to be adopted depends upon the travelling distance; the tractor-pulled scraper is favoured for short hauls, while the motorised version is now almost universally preferred where the size of the project permits.

EXCAVATING ACTION

The scraper is a self-loading, transporting and spreading machine predominantly used for general levelling. It is usual to cut on a downward gradient to take full advantage of gravity. For cutting, the bowl is lowered and the apron (1) opened, forward movement of the machine directs the cutting edge (2) into the soil causing it to boil upwards into the bowl. Excavation is carried out in layers of from 150 to 300 mm in depth, the levelling action is thus achieved as a gradual process. During the discharge stage ejection takes place whilst the unit is moving: again the height of the bowl is set to spread the material in a controlled layer and the soil is pushed out of the bowl with the aid of the ejector plate (3).

In general, tractor-pulled scraper capacities range from about 8 up to 30 m^3 heaped capacity, motorised scrapers from 15 up to 50 m^3 heaped capacity. The struck capacity is the volume of material contained in the bowl, when levelled off even with the top of the bowl. Some manufacturers quote the SAE (Society of Automotive Engineers, USA) heaped capacity of the bowl (see fig. 2.21), which represents the volume of material contained in the bowl allowing for heaping at a slope of 1:1.

Figure 5.1 Scraper action

TOWED SCRAPER

The towed scraper comprises a tractor, frequently this will be a bulldozer, and a towed bowl supported on two axles running on four wheels. Large volume tyres are needed to cope with the heavy loads to be transported over rough uneven surfaces. A powerful bulldozer, capable of providing 300 hp (223 kW) or more, fitted with specially strong and deep webbed tracks, is needed to provide the necessary traction during the loading operation. The loading cycle takes up to 2 min, with about 90% of the bowl filled in the first minute, but because of the slow travelling speed to the discharge area, output is considerably reduced. In practice, hauls greater than 300 m make the method uneconomic relative to other means of removing the material. However, outputs of the order of 40 m^3/h of struck material are achievable with, for example a 9 m^3 heaped capacity scraper. The main advantage over the motorised system is the ability to load in heavy soils, to load on the upwards gradient, to manoeuvre in a small turning circle and to 'pump' load in patches of hard material.

ejector plate

Figure 5.2 Towed scraper

MOTORISED SCRAPER

These machines were developed during the 1950s in an attempt to improve production on large earthmoving projects. Unlike the towed scraper, the engine is self-contained within the machine and power thus supplied directly to the wheels. The whole unit is supported on very large volume tyres.

The excavating and hauling action is carried out in a similar way to the towed scraper but frequently requires pushing assistance during the loading phase because of loss in traction when using wheels rather than tracks. However, haul speeds up to 60 km/h are possible with well-graded roads, yielding considerable improvements in output. For example, with the 20 m^3 heaped capacity scraper bowl output up to 150 m^3/h hauling over 1 km is possible.

The modern motorised scrapers are large earthmoving machines, ranging from 15 m^3 to 50 m^3 heaped capacity equipped with one or two large diesel engines capable of high power output, commonly up to 500 hp (512 kW) each. As with most modern excavation equipment the transmission is usually achieved through a torque converter and power-shift gears to the wheels. The gearbox often has from six to eight gear stages: the bottom four gears provide the high torque demanded during the loading operation and the top gears allow the high travelling speeds along the haul. Several types of scraper are available to suit different circumstances.

STANDARD SINGLE-ENGINE SCRAPER

The standard scraper (Fig. 5.3) comprises the scraper bowl (1) mounted on a single rear axle (2). The front end is hitched to the drive axle (3), by means of a single arm called, by virtue of its shape, the *swan neck* (4). The bowl height is controlled through a pivot attachment to the swan neck and hydraulic cylinders (5). The standard scraper is not self-loading and requires a pusher bull-dozer to provide the necessary traction.

Figure 5.3 Motorised scraper

DOUBLE-ENGINE SCRAPER

The double-engine scraper is similar to the standard unit in both its construction and operating action. However, the rimpull required on soils with high coefficients of resistance (i.e. well rutted, soft, etc.) and/or high uphill gradients is often best achieved with four-wheel drive to utilise the total available grip (i.e. traction). A second engine is then usually located over the rear axle (fig. 5.5).

Figure 5.4 Double-engine scraper

Figure 5.5 Four-wheeled tractor scraper

METHOD OF WORKING WITH THE MOTORISED SCRAPER

PUSH LOADING

The method is used mainly with the standard scraper. The bulldozer is fitted with a very robust blade and mounted on to the dozer frame through shock

(a) BACK TRACK LOADING

(b) CHAIN LOADING

Figure 5.6 Push loading systems

absorbers to take up impact, while the scraper is stiffened at the rear to form a push block. The pusher is used only during the loading phase and the scraper moves off under its own power as soon as the bowl is fully loaded. Generally a pusher serves from three to five scrapers working as a team, so although in practice high production is possible, the high cost of the scraper and bulldozer team limits the use of the motorised system to fairly large projects. Several push load systems are in use and are shown in fig. 5.6.

As a general rule:

$$\frac{\text{number of scrapers per pusher}}{} = \frac{\text{cycle duration of each scraper in minutes}}{\text{cycle duration of each pusher in minutes}}$$

Number of Rubber-tyred Motor Scrapers Serviced by a Bulldozer

It is assumed that a single dozer comprises the pusher set and that the average speed of a scraper is 30 km/h along a haul road. Two or more dozers per set may be required, depending upon the soil, size of scraper and dozer.

Haul distance (one way) (m)	Number of motor scrapers per pusher set
100	2
200	2
300	2
600	3
900	4
1200	5
1500	6
1800	7
2100	8
2400	8
3000	10

THE PUSH–PULL METHOD

This is an arrangement whereby two scrapers in turn assist each other to load without the need for a bulldozer pusher, however, a reasonable amount of co-ordination is necessary. The method is fairly effective in all but very heavy

Figure 5.7 Push–pull scraper connection

soils. Usually double-engine scrapers are used in tandem, whereby the rear scraper pushes the front scraper during loading and then the front machine pulls the rear scraper when it is loading. Clearly, the machines must be robust if they are to be operated in this way and modifications to the standard model include: strong swan neck and bowl hitch arms, a cushion push block and bail at the front end, and a hook and stiffened frame at the rear. A typical push–pull connection is shown in fig. 5.7 in which the mechanism is controlled hydraulically from the driver's cab.

Working Cycle

1. Scraper I ahead of scraper II after turning. Relative travelling speeds synchronised to within 2–4 km/h.
2. Scraper I begins to load with the help of a push from scraper II.
3. Scraper II loads with the help of a pull from the fully laden scraper I.
4. Connection link between scraper I and scraper II is broken.
5. Scraper I accelerates away, and scraper II follows slightly off line.
6. Scrapers discharge.
7. Scrapers turn.
8. Scraper II takes the lead.
9. Scrapers turn to continue the loading operations at (1) with scraper I regaining the lead.

Figure 5.8 Push–pull scraper system

The size of the bulldozer to use as a pusher is related to the carrying capacity of the scraper and depends upon the hardness of the soil to be loaded. More than a single dozer may be needed to provide sufficient power, in which case the dozers are referred to as the 'pusher set'. Typically the power of the pusher set to match a standard scraper for the push loading method is shown in table 5.1. Even the push–pull method often requires dozer assistance.

Table 5.1 Scraper and matching pusher dozers

Heaped capacity of scraper in yd³ (m³)	Single-engine scraper hp (kW)		Double-engine scraper hp (kW) Push-pull*
	Standard scraper	Pusher dozer set	
20(15.3)	350(261) up to	500(373)	450(335)
30(22.9)	450(335) up to	900(670)	700(521)
40(30.6)	550(410) up to	1200(894)	950(707)
50(38.3)	550(410) up to	1200(894)	950(707)

* In the push–pull system two scraper units are used.

SELF-LOADING SCRAPER

The scrapers described above are, in general, not self-loading because of the high resistance encountered when cutting into the soil. In an attempt to overcome this problem, the apron is removed and replaced by an elevator which aids the loading action and helps to offset the power required for traction purposes. Such machines are self-loading in all but very hard or heavy soils.

Figure 5.9 Elevating scraper

GENERAL POINTS TO CONSIDER WHEN USING SCRAPERS

1. Keep haul roads broad enough to accommodate overtaking.
2. Ensure that haul roads are well maintained with graders and free from deep rutting and grooving and well drained.
3. Construct hauls with gradients less than 5% if possible, max 12%.
4. Plan haul road routes to avoid bottlenecks, short lengths of sharp gradient and curves.
5. Try to keep the rolling resistance as low as possible.
6. Towed scrapers are economic up to 300 m haul distance.
7. Single-engine standard scrapers are suitable for light soils, and low rolling resistances along the haul and require pusher assistance, economic up to about 3 km haul.
8. Push–pull scrapers – must be double-engined and robust. Capable of similar work described above under (7).
9. An elevating scraper is 10–15% heavier than the equivalent standard scraper and generates a greater rolling resistance. Purchase price is higher but as it is a self-loading machine it may be more economic than the standard scraper if push loaders are not needed on the project.
10. The single-engine scraper has lower fuel costs than either the double-engine or elevating scraper.
11. Push loading gives a shorter loading time than push–pull loading.
12. Cuttings – start along the outer edges of the projected cutting and work downhill towards the fill area. Maintain the formation in a convex shape to lean the scraper towards the sides of the cutting and thereby encourage the maintenance of the full width of the cut. Continuously trim the cut sides with a grader (or a bulldozer as a second best), pushing the material to the scraper for removal. The sides should be cut in a series of steps by the scraper, as governed by the depth of soil layer attempted by the scraper.

13. Fillings – embankments are formed in layers working inwards from the outer limits. The formation is made concave to lean the scrapers towards the centre and so reduce the possibility of the scraper slipping over the edges.

PRODUCTION DATA

Ideal outputs from the towed and motorised scrapers are given in figs. 5.10 and 5.11 respectively. In practise, production may vary considerably from these recommendations and should be adjusted accordingly as typified by the example on output estimation in chapter 2.

HAUL ROADS

RESISTANCE TO MOVEMENT

The haul road is the length of track connecting the load and discharge areas on an excavation site. The haul road can quickly become rutted and holed with the constant pounding of the earthmoving plant. The production data given in figs. 5.10 and 5.11 assume a good haul road with a rolling resistance of about 2%. Rubber-tyred scrapers are particularly affected as the haul road deteriorates. For example, as the rolling resistance approaches 10%, output from a scraper could be halved, because the machine is forced to travel at a much lower speed and far more manoeuvring is involved. The rolling resist-

Figure 5.10 Ideal output of the towed scraper

Figure 5.11 Ideal output of the motorised scraper

Assumptions

(i)	Loading time	0.8 min
(ii)	Turning and dumping time	0.6 min
(iii)	Manoeuvring time	0.6 min
(iv)	Travel speed loaded	20 km/h
(v)	Travel speed empty	35 km/h
(vi)	Uninterrupted production	
(vii)	Rolling resistance of haul road less than 5%	

NOTE

(viii) Output volumes are bulked quantities, for in-place outputs reduce by an appropriate swell factor

ance is expressed as a percentage of the combined weight of the excavating machine and any self-contained load. Clearly the machine must supply sufficient power to the wheels to overcome rolling resistance. Table 5.2 gives approximate values of percentage rolling resistance for rubber tyres and also tracks operating on different soil types.

Table 5.2 Rolling resistance

Ground surface condition	Rolling resistance (%)	
	Rubber tyres	Tracks
Concrete	1.5	0
Fine gravel/sand	2.0	0
Loose gravel	10.0	5
Loose sand	10.0	5
Soft mud	16.0	7
Deeply rutted loam	16.0	7
Loose loam	4.0	3
Firm loam	2.0	1
Packed snow	2.5	0

RIMPULL AND TRACTION

The maximum speed attainable by a moving machine naturally depends on the rolling resistance of the ground, however, the grip (traction) between the ground surface and the wheels or tracks must be sufficient to prevent slippage and therefore must exceed the supplied rimpull, if the vehicle is to move at all.

The *supplied* rimpull is the tractive force applied between the tyres of the drive wheels and ground surface, measured in newtons (N) and is determined by the power of the engine and influenced by the efficiency of the transmission, the gearing ratios and wheel/track arrangements, and is further affected by the influence of altitude and air temperature on the engine power.

The *usable* rimpull, i.e. traction, however, depends upon the coefficient of traction for rubber tyres or tracks running on particular soils. The coefficient of traction is defined as the factor by which the load (i.e. downward force) on the drive wheels or tracks may be multiplied before slipping occurs. This value is the maximum *usable* rimpull, which clearly must be greater than the supplied rimpull or the wheels will spin.

EXAMPLE

A rubber-tyred motorised scraper has a capacity of $21/30$ yd^3 ($16/23$ m^3) (struck/heaped). The machine is operated on firm loam up a gradient of 6% at an altitude of 900 m where the air temperature during the day is 30 °C. The rated power of the engine is 450 hp (335 kW). The gross laden weight of the unit is 72 000 kg, made up of 36 000 kg self-weight and 36 000 kg payload. The total load is distributed with 50% on each axle. Determine:

(a) if the wheels will slip;
(b) if sufficient rimpull is available given the travelling speeds for the following gear ratios:

1st gear	5 km/h
2nd gear	12 km/h
3rd gear	18 km/h
top gear	40 km/h.

Information

Approximate coefficients of traction are given in table 5.3. For wet materials the coefficients are difficult to determine but cohesive materials are especially affected and the coefficient of traction could be reduced by up to 50%.

Table 5.3 Approximate coefficients of traction

Type of surface (dry)	Rubber tyres	Tracks
Smooth concrete	0.8–1.0	0.3–0.6
Clay	0.5–0.8	0.6–0.9
Firm sand/gravel	0.3–0.8	0.7–0.9
Loose sand	0.1–0.2	0.3–0.5
Loose gravel	0.2–0.4	0.4–0.7
Packed snow	0.1–0.4	0.2–0.6
Firm loam	0.4–0.8	0.6–1.0
Loose loam	0.4–0.6	0.7–1.0

Solution

(a) (i) From table 5.3, the coefficient of traction for firm loam is, say, 0.6. The gross laden weight of the unit is 72 000 kg of which 36 000 kg is supported on the drive axles. Therefore the maximum usable rimpull is $36\,000 \times 0.6 \times g$ N, where g is the force of gravity, 9.81 m/s^2, which reduces to 212 kN.

(ii) Maximum power supplied from the engine $= 450 \times 0.746 = 335$ kW. Typically for this size of machine, the efficiency in transferring engine power into rimpull is about 80% and supplied rimpull (in kN)

$$= \frac{3.6 \times \text{engine power (kW)} \times \text{efficiency}}{\text{speed (km/h)}}$$

$$\text{Supplied rimpull: in 1st gear} = \frac{3.6 \times 335 \times 0.8}{5} = 193 \text{ kN}$$

$$\text{Supplied rimpull: in 2nd gear} = \frac{3.6 \times 335 \times 0.8}{12} = 80 \text{ kN}$$

$$\text{Supplied rimpull: in 3rd gear} = \frac{3.6 \times 335 \times 0.8}{18} = 54 \text{ kN}$$

$$\text{Supplied rimpull: in top gear} = \frac{3.6 \times 335 \times 0.8}{40} = 24 \text{ kN}$$

(iii) *Approximate reduction in engine horse power for altitude* – depends upon the manufacturers' engine performance specification but far four-cycle diesel engine the losses are approximately:

Altitude
0–300 m ....no loss in performance
300 m and above ....reduces available rimpull by 3% per 300 m.

(iv) *Approximate reduction in engine power for temperature*

Temperature (°C)	15°	0°	15°	30°	45°
Reduction (%)	+6	+3	0	−3	−5

Therefore total reduction of rimpull $= 6\% + 3\% = 9\%$.
Available rimpulls corrected for temperature and altitude:

1st gear $= 176$ kN
2nd gear $= 73$ kN
3rd gear $= 49$ kN
top gear $= 22$ kN.

Supplied rimpull in first gear 176 kN. This is less than usable rimpull of 212 kN (see (i)), therefore the wheels will not slip.

(b) (i) Rolling resistance from table 5.2 for firm loam $= 2\%$.
Supplied rimpull required to overcome rolling resistance
$= $ total weight of unit $\times$ rolling resistance

$$= 72\,000 \times \frac{2}{100} \times \frac{9.81}{1000} = 14 \text{ kN}$$

(ii) Supplied rimpull required to overcome grade resistance

$$= 72\,000 \times \frac{6}{100} \times \frac{9.81}{1000} = 42 \text{ kN}$$

Therefore total supplied rimpull required = 56 kN.

The scraper will generate sufficient rimpull in second gear to climb the grade and thus if the unit is to be operated at heaped capacity the maximum speed along the haul is 12 km/h. However, top gear may be engaged where there is no gradient. Should it be necessary to operate the machine on loading in first gear, on an upward gradient, then clearly the available rimpull is 176 − 56 = 120 kN. In effect this value is the *draw bar pull*, and is the available pull which could be exerted on a load that is being towed, for example a box scraper attached to a crawler dozer.

Note: The reader should now repeat the calculation in the unloaded condition.

EXAMPLE OF A MACHINE SELECTION EXERCISE

Mass Haul

Bulk earthworks operations on a road construction contract involve forming cuttings and embankments. All earthmoving is to be carried out by tractor-pulled scrapers. Tips and borrow pits are available as required.

Table 5.4 gives quantities computed from longitudinal sections, and cross sections. Bulking is 10% after recompaction. Costs are as follows:

(i) dig and form banks up to 200 m haul, £0.30 per m³;
(ii) dig and form banks between 200 m and 300 m, £0.80 per m³;
(iii) dig and form banks between 300 m and 400 m, £1.60 per m³;
(iv) cart to tip, £0.50 per m³;
(v) bring from borrow pit, £0.75 per m³.

Solution

Table 5.4

Length along road (m)	Volume (m³)			Cumulative volume (m³)
	Cut	Fill	Fill $\times \frac{10}{11}$	
0	0			0
200	40 000			40 000
400	70 000			110 000
600	30 000			140 000
800		20 000	18 200	121 800
1000		40 000	36 400	85 400
1200		50 000	45 500	39 900
1400		20 000	18 200	21 700
1600	40 000			61 700
1800	40 000			101 700
2000		20 000	18 200	83 500

Figure 5.12 Mass
haul diagram

The cumulative volumes are plotted graphically on fig. 5.12. With the aid of dividers the 200 m, 300 m and 400 m hauls are marked on the diagram and the respective volumes of earth moved by the scraper are obtained by subtracting values from the gradient turning points. The volumes of cut and fill are read off the graph between the respective values established between the haul limits.

The cost for each combination of haul, cut and fill are calculated as follows:

200 m haul

Hauled	= 37 000 m³	cost	= 37 000 × 0.3	=	£11 100
Fill	= 94 000 + 5500	cost	= 99 500 × 0.75	=	£74 625
Cut	= 128 000 + 55 000	cost	= 183 000 × 0.5	=	£91 500
					£177 225

300 m haul

(ii) Hauled 300 m = 53 000 m³
(i) Hauled 200 m = 37 000 m³
                                  16 000 m³

Fill	= 83 500 m³	(i)	cost	= 37 000 × 0.3	=	£11 100
Cut	= 167 000 m³	(ii)	cost	= 16 000 × 0.8	=	£12 800
			cost	= 83 500 × 0.75	=	£62 625
			cost	= 167 000 × 0.5	=	£83 500
						£170 025

400 m haul
Hauled 400 m = 70 500 m^3
Hauled 300 m = 53 000 m^3
Hauled 200 m = 37 000 m^3

Fill	= 66 000 m^3	(i) cost	=	37 000 × 0.3	= £11 100
Cut	= 149 500 m^3	(ii) cost	=	16 000 × 0.8	= £12 800
		(iii) cost	=	17 500 × 1.6	= £28 000
		cost	=	66 000 × 0.75	= £49 500
		cost	=	149 500 × 0.5	= £74 750
					£176 150

The cost of each respective haul distance is plotted on fig. 5.13, where it is seen that the economic haul distance is approximately 300 m.

Figure 5.13 Economic haul distance = 300 m (i.e. scrapers are only used up to 300 m hauls; outside this limit material is transported to tip or brought in from borrow pits)

Chapter 6   MOVING EXCAVATING MACHINES – THE GRADER

INTRODUCTION

Many earthmoving projects require the final ground to be accurately finished, so that the surface is smooth and level without undulations and ridges. Although a skilful driver using a bulldozer can in many instances achieve adequate results, the grader (fig. 6.1) has been specifically developed for trimming the subgrade, sub-base surface on roads and road cuttings and banks, for smoothing off the walls on earth-fill dams and maintaining haul roads. It is self-transporting and supported on two or three axles. The large road programmes over the past few years have demanded heavier and more robust machines and the three-axle configuration has proved necessary to carry the large engine and provide more traction and is today the more popular type.

MOULDBOARD

Figure 6.1 Grader

MECHANICAL ARRANGEMENT

Figure 6.2 Leaning to resist side load

The engine, transmission, driver's cab and controls are supported on the rear frame. Modern machines allow the driver to work in the seated position who until fairly recently had to stand in order to see clearly to operate the mouldboard.

The front wheels support a long bridging beam from which the mould-board is hung. On some types the beam is pivot-jointed to the rear frame to allow a reduced turning circle, increased manoeuvrability and to permit offset axle grading. On others the connection is rigid and steering is only possible through the front axle. This design allows the wheels to: (a) lean (fig. 6.2), up to about 15° each side of vertical to resist side loads, for example, when angling the mouldboard, and (b) to operate at different levels (fig. 6.3) while shaping banks, making ditch cuts, forming crossfalls and so on. Combination of the two facilities means that direction may be held without the need for excessive concentration by the driver on steering, so freeing his attention for direction of the mouldboard.

Figure 6.3 Tilting axle

THE MOULDBOARD

The mouldboard on modern machines is operated hydraulically from the driver's cab, and although it works similarly to the bulldozer blade, it is hung between the axles, and the magnifying effect of uneven and stony ground is very much reduced, as demonstrated in fig. 6.4. The pitch of the mouldboard is adjusted to suit the work. It is tilted back when cutting and forwards when spreading. The more upright the mouldboard, the greater the mixing and rolling which will be given to the material being spread.

Figure 6.4 Comparison of action between the grader and bulldozer

GRADER BLADE LENGTH AND CORRESPONDING ENGINE POWER

Engine		Blade length	Machine weight
hp	(kW)	(m)	(kg)
35–50	(26–37)	2.5	2 500
50–70	(37–52)	3.0	3 500
70–90	(52–67)	3.0–3.5	6 500
90–120	(67–89)	3.0–3.5	8 000
120–140	(89–104)	3.5–4.0	10 000
140–160	(104–119)	3.5–4.0	13 000
160–180	(119–134)	3.5–4.0	14 000
180–300	(134–224)	4.0–4.5	25 000

METHODS OF WORKING WITH THE GRADER

The blade can be used in several positions for:

(a) Levelling and trimming on the horizontal, with the mouldboard central or swung out either to the left or right (fig. 6.5a). If the mouldboard is set at an angle on plan, the material will roll off the blade to form a windrow. However, with the blade at right angles to the line of movement, then only spreading or trimming is obtained.

(b) Levelling and trimming to the slope and vertical face (fig. 6.5b).

(c) Forming ditches (fig. 6.5c). The mouldboard is angled both on plan and in the vertical and set such that the blade just protrudes beyond the outside line of the wheels nearest to the ditch to be shaped. A windrow is formed along the top of the ditch. The ditch is deepened gradually, trimming off a layer at a time, keeping the nearside wheels in the ditch.

(d) Backfilling along trenches (fig. 6.5d). The action is similar to that for producing a windrow.

(a)

(c)

(b)

(d)

Figure 6.5 Methods of working with the grader

CONTROLLING THE BLADE

Without guidance the blade trims to the plane dictated by the unevenness of the ground, but because the mouldboard can be positioned by means of

Figure 6.6 Wire and
sensors to control
grading levels

hydraulic cylinders, the driver is able to superimpose some measure of control, independent of the posture taken by the wheels. However, in order to obtain very smooth and even surfaces it is necessary to use some form of predetermined levelling device. Frequently for road works, wires are set along the direction line and sensors attached to the mouldboard adjust the blade height automatically. As an alternative device, a laser beam activates photoelectric cells which control the hydraulic adjusters fitted between the blade and the frame of the grader.

EXTRA ATTACHMENTS

Many graders also carry a scarifier mounted just in front of the mouldboard, it is raised and lowered hydraulically and used to break up the soil to facilitate an easier grading action. A ripping tool may sometimes be mounted at the rear end and a conventional bulldozer blade at the front.

RECOMMENDED OPERATING SPEEDS

	km/h
Grading site roads	4–9
Scarifying (e.g. soil stabilisation)	8–18
Forming ditches	4–8
Spreading	4–10
Trimming and levelling	9–40
Snow ploughing	8–20
Self-transporting	10–40

The demands on the grader clearly are very varied, so robust engines up to 300 hp (223 kW) are common. Hydraulic control of the attachments is standard on the latest models.

Chapter 7   MOVING EXCAVATING MACHINES – THE LOADER SHOVEL

The loader shovel, sometimes called the tractor shovel or front end loader, exists in two forms; with tracks (fig. 7.1) or wheels (fig. 7.2). It is a machine which serves the purpose of both the fixed-position excavator and transporter over short distances of perhaps 10–20 m.

It is the ability of the machine to cope with duties usually associated with the face shovel and also to overlap with the functions of the bulldozer which has caused the rapid increase in numbers in plant fleets over recent years.

Excavating from a level surface

Excavating against sloping surface

Transporting

Loading

Figure 7.1 Tracked loader shovel

Figure 7.2 Wheeled
loader shovel

The choice between tracks or wheels is fairly straightforward. The tracked loader is a genuine excavator, whereas the wheeled version is more suited to stockpiling and digging in loose soils. The tracked machine can apply more traction and will cope with tougher conditions. The tracked loader is typically available with bucket sizes of up to 9 yd^3 (7 m^3), whilst buckets over 40 yd^3 (30 m^3) are possible, especially on the larger wheeled machines. The tracked loader is of course limited by its slow travel speed and becomes uneconomic when the distance from the excavation or discharge points exceeds about 80 m. The wheeled loader being faster is viable up to 200 m or so, and is able to travel on tarmacadam roads without causing damage and is certainly more mobile and manoeuvrable than the equivalent tracked machine. However, tracks are more stable on soft and rutted surfaces, which helps to improve the production performance.

MECHANICAL ARRANGEMENT OF THE TRACKED LOADER

Tracked loaders are almost identical in appearance to bulldozers, the engine, gearing and hydraulics are all similar (fig. 7.3) but the tracks are positioned more to the front to provide counterweight when loading and to distribute the bearing pressure more evenly when the bucket is fully loaded. Unlike the bulldozer, however, independent and differential movement of the tracks is impossible.

LOADER FRAME

FRONT FRAME

Figure 7.3
Construction of the
tracked loader

TRACK FRAMES JOINED BY
RIGID BARS

FRAME ASSEMBLY

The bucket attachment is controlled by hydraulic rams and an automatic cut-out device disengages the hoist control lever when the bucket reaches a predetermined height. A similar device is sometimes fitted on the tilt rams to keep the bucket horizontal when tipping.

MECHANICAL ARRANGEMENT OF THE WHEELED LOADER

Figure 7.4 Pivot connection

Figure 7.5 Rear axle pivot action

Structurally the wheeled loader is quite different from the tracked machine and has in fact been developed from the farm tractor into a form which today consists of a pivoted frame with the engine mounted over the rear wheels (fig. 7.4) and the cab over either the front or rear frame dependent upon the manufacturer's preference. The pivot arrangement gives the loader very good manoeuvring capabilities, facilitating the front frame to turn in plan up to 40° either side from the forward position (fig. 7.2), to work in a relatively small turning circle.

Power is supplied from a diesel engine through a torque converter, and power-shift gears to drive the wheels; four-wheel drive is common, necessitating all wheels to be the same size. The machine, however, can be operated as two-wheel drive only. The rear-wheel drive loader is better at digging while front-wheel drive gives better traction when carrying a full bucket. On most machines the front axle is fixed, whilst the rear axle may oscillate up to ±15° from horizontal (fig. 7.5), i.e. a total of 30° to incorporate differential movement between the wheels and allow more effective action on uneven ground.

MODERN DEVELOPMENTS IN LOADER DESIGN

Some very modern machines have complete hydrostatic drive, whereby displacement pumps, powered direct from the engine, drive displacement motors on the wheels or tracks. Furthermore, a few of the wheeled machines have a hydraulic rather than a pivot connection between the two frames to give articulation in the vertical plane for both the rear and front axles.

METHODS OF WORKING WITH THE LOADER

The size of the machine depends upon the particular circumstances of the job. Both the tracked and wheeled types are able to carry out similar operations and are basically used for loading loose material or for excavating at the level of the wheels or tracks in fairly loose soils. By careful positioning of ramps the excavator may be used to form large pits and other deep excavations. The machine is also very suitable for stripping soil in thin layers – acting like a bulldozer.

Modern machines are quite robust and can compete favourably with face shovels, especially in quarries.

LOADING METHODS

(a) *V-loading* (fig. 7.6a) is a very efficient method and is most frequently adopted when the wagons are able to take up the required positions.

(b) *Side dump loading* (fig. 7.6b) requires no turning of the loader and is used mainly for backfilling ditches or close quarter loading. It is not commonly used in general construction work.

(c) *Cross loading* (fig. 7.6c) requires far more manoeuvring than the V-method and is slightly less efficient.

(d) *Straight loading* (fig. 7.6d) is used when it is possible to place the wagon along the line of the excavation, the method appears in principle to be very effective. However, the need to co-ordinate two machines is a little cumbersome and therefore seldom used.

(e) *Overhead loading* (fig. 7.6c) is used mainly in tunnelling work.

(a.) V LOADING

(c.) CROSS LOADING

(b.) SIDE DUMP LOADING

(d.)STRAIGHT LOADING

(e.)OVERHEAD LOADING

Figure 7.6 Loader movements

THE LOADING BUCKET

(a) *The general purpose bucket* (fig. 7.7a) is made from thick high-strength, wear-resilient plate. The cutting edge extends forward, and the addition of steel teeth gives an improved penetration.

(b) *The rock bucket* is made of special heat-treated alloy steel with high strength and wear resistance. The bucket has either a V-cutting edge (fig. 7.7b) or a modified V-edge (fig. 7.7c) and cut-back sides to aid penetration.

Figure 7.7 Loader buckets

(a) GENERAL PURPOSE BUCKET   (b) ROCK BUCKET   (c) ROCK BUCKET

(c) *The side dump bucket* (fig. 7.8) dumps to the left or to the right side, providing considerable advantages in confined areas and for backfilling along trenches.

(d) *The multipurpose bucket:* a four-in-one bucket (fig. 7.9) was marketed originally by the Drott Company and since then any machine fitted with this type of bucket has tended to be called a 'Drott'. The bucket action is hydraulically controlled and used in one of four forms; a dozer blade, shovel, grab or clamshell, scraper. The most popular size is about 1 m^3 for use with small and medium-sized bulldozers and wheeled or tracked loaders. It is this bucket attachment which has provided the loader with most of its advantages over alternative types of excavator.

Figure 7.8 Side dump bucket

(e) Other attachments include log grapples, fore-lift tool carrier, brush, etc.

Figure 7.9 Four-in-one bucket

DOZER   SHOVEL   CLAMSHELL   SCRAPER

WHEELED BACKHOE/BUCKET LOADER

Many construction sites require excavation for manholes, drain trenches, etc., and the loading of loose material into wagons to be carted away to tip. Much of this work, particularly for building-type projects, is on a small scale and takes place spasmodically. The backacter-type loader shovel adequately fills such demands. The machine comprises a rigid frame (no pivoting is possible), a loader bucket and backhoe. Only the rear wheels are powered, so the front wheels do not need to be of the same size. The engine and transmission are in general similar to those of other loading units and operate with a similar loading action, however, the driver usually moves to a separate seat and control panel to operate the backhoe which is mounted on a horizontal bar for digging wide trenches. The backhoe can turn through 90° in plan and excavate approximately 5 m below the wheels. The machine must be stabilised by jacks in this mode, otherwise too much movement and strain on the axles will be generated from the large voluminous rear tyres.

Figure 7.10 Wheeled backacter/bucket loader

Figure 7.11 Skid-steer loader

SKID–STEER LOADER

The machine is wheel mounted and very robust. Hydrostatic drive facilitates quick forward and backward changes of direction to provide excellent manoeuvrability in confined spaces such as basement excavation and inside cofferdam enclosures. Typically the weight range of models varies from about 1000 kg up to 6500 kg equipped with buckets of from 0.3 to 2.5 m³ capacity, respectively.

LOADERS – POPULAR RANGE

Bucket heaped capacity yd³	(m³)	Engine hp	(kW)
1–1½	(0.76–1.14)	60–80	(45–60)
1½–2¼	(1.14–1.72)	80–100	(60–75)
2¼–3	(1.72–2.30)	100–150	(75–112)
3–4	(2.30–3.06)	150–220	(112–164)
4–5½	(3.06–4.21)	220–320	(164–238)
5½–7	(4.21–5.36)	320–400	(238–298)
7–9	(5.25–6.75)	400–700	(298–525)

PRODUCTION DATA

The production information given in figs. 7.12 and 7.13 is for ideal conditions and should be adjusted for a particular construction site, as explained in chapter 2.

EXAMPLE OF MACHINE SELECTION

Select suitable plant to excavate one million cubic metres of firm silty clay for the foundations of a modern power station.

Information

(a) All material may be directly loaded into dump trucks and ramped up to ground level.
(b) Average haul for excavator is 30 m.
(c) Swell factor is 1.2.
(d) Excavating season, March to October inclusive.
(e) Tip located on site.

Figure 7.12 Ideal output of the tracked loader shovel

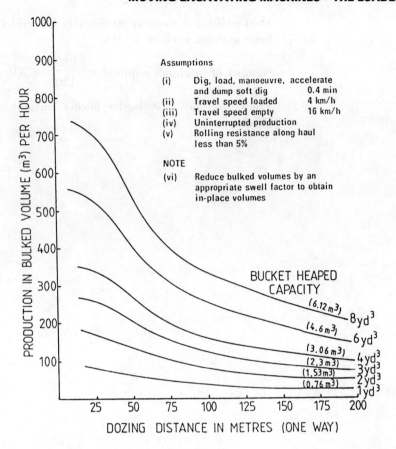

Figure 7.13 Ideal output of the wheeled loader shovel

Assumptions

(i)   Dig, load, manoeuvre, accelerate
      and dump soft dig   0.4 min
(ii) Travel speed loaded   4 km/h
(iii) Travel speed empty   16 km/h
(iv) Uninterrupted production
(v)   Rolling resistance along haul
      less than 5%

NOTE

(vi) Reduce bulked volumes by an
      appropriate swell factor to obtain
      in-place volumes

BUCKET HEAPED CAPACITY

(6.12 m³) 8 yd³
(4.6 m³) 6 yd³
(3.06 m³) 4 yd³
(2.3 m³) 3 yd³
(1.53 m³) 2 yd³
(0.76 m³) 1 yd³

PRODUCTION IN BULKED VOLUME (m³) PER HOUR

DOZING DISTANCE IN METRES (ONE WAY)

Solution

In the dry season trucks may approach the excavation area on firm ground, therefore a tracked vehicle is selected, because of its improved traction capabilities over wheeled loaders.

Output from a 4 yd³ (3.06 m³) heaped capacity tracked loader given in fig. 7.12 is 200 m³/h. This is a bulked production rate.

$$\text{In-place output} = \frac{\text{bulked output}}{\text{swell factor}} = \frac{200}{1.2} = 167 \text{ m}^3/\text{h}$$

Potential losses (%):

Weather	10
Breakdown	5
Operator efficiency	7
Waiting	10
	32

$$\text{Realisable in-place production rate} = 167 \times \frac{68}{100} = 113 \text{ m}^3/\text{h}$$

$$\text{Time required} = \frac{1\ 000\ 000}{113} = 8850 \text{ h}$$

Muck-shifting season, approximately 30 weeks with an average 50 hour working week = 1500 h.

Number of machines required $= \dfrac{8850}{1500} = 5.9$, say 6 excavators

Truck size, say 5 × loader bucket capacity = 5 × 4 = 20 yd^3 (15.3 m^3).

Chapter 8   TYRES

Tyres are required to spread the machine weight on to the ground surface, to convert the rimpull at the points of contact with the ground surface into traction and to aid and improve the steering capabilities of the machine.

TYRE CONSTRUCTION

The design and manufacture of tyres has improved slowly during the past 10 years with new versions to match the improved performance of the present generation of machine only gradually appearing from the major manufacturers. Currently there are two principal methods of tyre construction; *cross-ply* and *radial ply*, both types are mounted on earthmoving machines (but not mixed together). A special *beadless* tyre is now being introduced on some loader machines.

CROSS-PLY TYRE

The cross-ply tyre is shown in fig. 8.1 and consists of:

Figure 8.1 Cross-ply tyre

1. Tread – this is the outer wearing surface which makes contact with the ground and gives protection to the inner layers. The tread is an important part of the tyre, since the rate of wear and friction coefficient are influenced by the quality of the rubber and tread pattern.
2. Side walls – these are a protective layer of rubber covering the tyre plies.
3. Belts – these provide an intermediate layer of protection, usually confined to the tread area. They wrap around the tyre circumference.
4. Body plies – consist of alternating layers of synthetic material which cross each other at a 45° angle. They form the main carcase of the tyre and provide its strength.
5. Inner liner – this is an inner lining which seals in the air.
6. Beads – these consist of bands of steel wire integrated with the body plies to anchor the tyres to the wheel rim. When the tyre is inflated, a good air seal is maintained.

RADIAL TYRE

The radial tyre is constructed in a different manner to the cross-ply tyre. The beads are made from a single bundle of steel wires, while the tyre carcase is a single layer of steel cables fixed radially from bead to bead to form an arch shape. A more flexible and wider bodied tyre is the result.

TYRE SIZE

The tyre size is usually stated, for example as 40–39, where the first number is the cross section width of the tyre and the rim diameter is the second number in inches. The tyre may also be designated by its aspect ratio, which is the ratio of the tyre cross section height to width. Aspect ratios are: standard tyre 1.0, wide tyre = 0.85 and low profile tyre = 0.65.

Thus a tyre designated 65R40–39 represents a low profile, radial tyre.

PLY-RATING

This is an index defining the strength of the tyre, and is directly related to the type of material forming the body plies. The ply-rating for a given tyre size may vary depending upon the working requirements.

For example for a tyre size 18–25 in:

Ply-rating	Permissible load on tyre (kg)	Tyre pressure N/mm^2	(atm)
12	4675	0.2	(2.0)
16	5675	0.275	(2.75)
20	6475	0.3	(3.0)
24	7275	0.425	(4.25)
28	8000	0.5	(5.0)

TYRE SELECTION

Generally, the larger the tyre the greater the machine output because large diameter tyres require lower tyre pressures and therefore present a larger contact area to the ground surface, so reducing rolling resistance and thus reducing tyre slippage and wear and tear. As a consequence treads may be shallower for the same tyre life, with a corresponding further reduction of rolling resistance. The introduction of the wide-bodied tyre, further emphasises these advantages. However, increased tyre diameter lowers the rimpull.

Presently, radial tyres are preferred for hauling operations because the heat build up tends to be less than with cross-plies and they have a slightly lower rolling resistance.

Finally, the lower the ply-rating the lower the heat build up. Use deep treads for better grip on soft ground and durability on rock or hard surfaces. Smooth or ribbed tyres are recommended for light flowing soils.

TYRE LIFE

The tyre life is affected by the tyre size and type, heat build up, travel surface, tyre pressure and load, and is further influenced by the operator's skill and vehicle maintenance. In particular, the tyre should be matched with the load to be carried and the correct tyre pressure maintained. As a further precaution manufacturers have introduced a system to relate heat build up and tyre safety known as the ton – mile/h ratings (= mean tyre load × average speed).

The mean tyre load is the weight on the tyre when the machine is fully loaded plus the weight on the tyre when the machine is empty, divided by two.

The average speed is calculated from the total distance travelled in kilometres during the work day, divided by the length of the work day in hours.

When operating scrapers, for example, the cost of tyres can amount to up to 30% of the total owning and operating costs of the machine. It is therefore important to observe the manufacturer's advice.

A GUIDE TO TYRE LIFE

| Machine | Operating conditions | | |
	Favourable (h)	Average (h)	Unfavourable (h)
Scrapers			
Twin engine	4000	3000	2500
Single engine, tractor	4000	3000	2500
Single engine, box	5000	3500	2500
Loaders	3500	2500	1500
Graders	5000	3000	2000
Trucks	4000	3000	2000

The tyre life on loaders can be improved by fitting chains.

Chapter 9   SPECIALISED EXCAVATING MACHINES

The rapid build up of the oil and gas industries during the past 40 years has called for pipelines in the distribution network to be buried below ground. As a result machines have been developed to improve the efficiency of excavating long sections of trench. In conjunction with this development, there has been the need for machines to cut canals and work vast mineral deposits.

The equipment for performing these duties most effectively is either the chained-bucket or rotary-bucket excavator. Both are expensive, and technically sophisticated machines, which require relatively large amounts of capital investment compared to most other types of earthmoving equipment.

CHAINED-BUCKET EXCAVATOR

The machine is very efficient for excavating shallow trenches in fairly loose soils, and the example shown in Fig. 9.1 is a simple type for use on small scale operations. Large and more sophisticated machines are available for more substantial work, such as canal construction.

chained bucket

Figure 9.1 Chained-bucket trenching machine

METHOD OF WORKING

The soil is excavated along the direction of travel by means of an endless chain of buckets mounted on a boom attachment fixed to the main frame. The contents are discharged on to a conveyor belt integral with the machine and then deposited into some other means of spoil disposal. Wear and tear tends to be rapid in abrasive soils as the drag force puts considerable tension into the chain, demanding almost continuous maintenance, with the associated high costs. Power is supplied either by a diesel or electric engine, the latter source being used mainly for large units. The range of sizes on the market is now sufficient to meet most demands, from the small unit of 20 m³/h output up to the giants of 300 m³/h. However, the length of trench must be long in order to justify the large capital purchase and would only be possible if much work of a similar character could be obtained in future contracts. Furthermore, the type of soil very much influences the output capability of the machine and heavy clay soils are almost impossible to excavate using this method. Thus if the nature of the soil is likely to change frequently over a run of trench excavation, the use of other traditional excavating methods on some sections may be necessary and the machine's economic advantage becomes seriously eroded.

ROTARY-BUCKET EXCAVATOR

Only on extremely large excavation works, for example opencast ore mining, does it become economic to use the large-wheeled bucket excavator. However, the rapid developments now taking place in design may result in the smaller types being operated for more general construction work. Currently machines with outputs of the order of 100 m³/h up to 20 000 m³/h are available.

A large-diameter wheel placed at the end of a long boom is rotated in the excavation face. The wheel consists of six to eight buckets arranged in a regular pattern around the rim, the diameter of which may vary from 2 m up to 20 m depending upon the size of the machine, buckets on the smaller units may be quite small or perhaps 10 litre capacity, increasing to 4000 litre on the giant versions. The wheel is rotated at 10–30 cycles/p min and about 25% of the circumference comes into contact with excavation face.

Figure 9.2 Rotary-bucket excavating machine

BUCKET SHAPE

THE CELLULAR WHEEL

The cellular wheel comprises an integral series of bucket cells through which excavated material is discharged on to a conveyor belt. Some difficulty may be experienced in dislodging cohesive soils from the cells, the only recourse then is to run the wheel at a lower speed to reduce the centrifugal force, a correspondingly lower output is obtained.

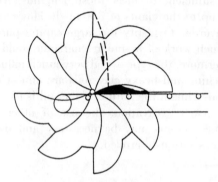

Figure 9.3 Cellular wheel

THE BUCKET WHEEL

The bucket wheel consists of single cells mounted on the rim, through which the soil falls as each bucket reaches its highest position on the rotation cycle. A ring plate prevents discharge taking place earlier. The arrangement is much lighter than that of the cellular wheel and also less liable to clog up in cohesive soils.

Figure 9.4 Bucket wheel

CUTTING ACTION

(a) Forward cutting (fig. 9.5) – the cells are eased into contact with the soil starting at (1) gradually deepening as the wheel rotates into the face. The upward rotation reacts against the weight of the machine and so provides an ideal method for tackling heavy soils.

Figure 9.5 Cutting action of a rotary-bucket excavator

(b) Downward cutting (fig. 9.6) – this time the cutting action finishes at the toe of the cutting (1) and a much cleaner surface is obtained. The applied force between cells and soil face is dependent upon the weight of the machine, little reaction can be gained from the soil in comparison with the forward cutting action.

Figure 9.6 Downward cutting

BLOCK AND LINE CUTTING

To excavate a face deeper than the diameter of the wheel, it is usual to remove the material in stages (fig. 9.7). In this way perhaps 20% of the potential output is lost, due to the forwards and backwards shunting action demanded of the machine. The alternative of leaving a deep under-reemed face would be rather unsafe.

$$H = \tfrac{2}{3} D$$

Figure 9.7 Cutting a deep excavation face

The block cutting system shown in fig. 9.8 or the alternative line cutting method in fig. 9.9 have been developed to deal with a deep excavation face.

Figure 9.8 Block
cutting method

Figure 9.9 Line
cutting method

PRODUCTION DATA

The output of the machine very much depends upon the type of soil and the
quantity which is available for continuous excavation. For estimating
purposes the following figures as used by Pohle at Aachen. T.H., Germany,
give a guide:

Wheel diameter (m)	4	6	8	20
Bucket capacity (m^3)	0.2	0.3	0.5	5
Soil – light (m^3/h)	400	800	1600	16 000
Soil – medium (m^3/h)	300	600	1200	12 000
Soil – heavy (m^3/h)	200	400	800	8 000

Chapter 10   DREDGES

Dredging is a highly specialised excavating method used predominantly for deepening rivers, estuaries and coastal areas for improved navigation; land reclamation as typified by the work of the Dutch in the Polder regions of Holland; the winning of sands and aggregates from river and sea beds; and specialist excavation on civil engineering works where the dredging barge can approach the site. The dredging action may be: (a) of the suction form which stirs up material and pumps the mixture on to nearby land, into barges or is stored on board, or alternatively (b) mechanical whereby the material is physically excavated and loaded into tugs and barges. Several different types of dredge are operated within these two broad classifications each with features to provide an advantage in a particular dredging situation.

SUCTION DREDGE

The suction dredge is basically a floating platform equipped with a powerful centrifugal pump and suction pipe. The material is first loosened by jetting in front of the suction pipe entrance using special high-pressure pumps, the resulting spoil and water mixture is then drawn up the suction pipe and pumped into barges or directly to shore. The exact output depends upon the length of pipe, suction pressure and friction loss. The pumped mixture consists of between 1:6 and 1:10 soil to water, and, on some of the larger vessels, over 10 000 m³/h may be pumped.

The method is most suited for operation in sands for land reclamation use. Where the site is too distant for the capacity of the pumps on the dredge the spoil is discharged into specially equipped barges then pumped away over distances of 10 km or more in large-diameter pipes. Positioning is achieved by winching from anchors.

Figure 10.1 Suction dredge

TRAILER SUCTION HOPPER DREDGE

Figure 10.2 Self-propelled hopper dredge

The hopper dredge is self-propelled and stores the excavated material in hoppers built into the hull and discharges through valves at the bottom of the vessel.

Suction pipes are mounted along the sides and may be pulled up for transporting. During dredging the boat is propelled forward at about 1 m/s (4 km/h) drawing the spoil up the large-diameter suction pipe through a draghead (fig. 10.3) which, being heavy and wide, presses into the bed resulting in a stirred up mixture of spoil and water which is then discharged into the hopper.

The largest types are capable of dredging to depths of 30 m and will store up to 10 000 m³ of spoil on board and work in waters with wave heights up to 2.5 m.

The output depends upon the soil type but dragheads are available for use in fine and coarse sands, clay and mud. Naturally, the volume of soil dredged will vary since the spoil is a mixture of soil and water, but modern pumps are capable of filling the 10 000 m³ hopper in 1 hour. The hopper dredge is a self-contained seagoing ship and can sail to almost any dredging location. It is used mainly in fine loose soils for improving waterway schemes.

Figure 10.3 Draghead

CUTTING SUCTION DREDGE

Figure 10.4 Cutterhead dredge

Where the effect of normal suction pressures or jetting is insufficient to loosen the soil, a mechanical method, such as the cutterhead dredge, is preferred. The equipment consists of a rotating head (fig. 10.5) fixed to the end of a stiff jib on which the drive motor and suction pipe are also carried. The cutter operates by rotating the arm in an arc and the dredged materials are pumped

Figure 10.5 Cutterhead

Figure 10.6 Walking the barge

ashore along a floating pipeline. Depths up to 15 m can be worked: the method is suitable for cutting soft to medium hard rock but is also capable of operating in clays, sands and granular deposits.

The cutting dredge is generally not self-propelled since the vessel needs to be held in position whilst working and this is achieved with two piles at the rear of the barge. Gradual positioning adjustment may be effected in conjunction with a system of anchors and ropes (fig. 10.6) and the barge is 'walked' forward by raising one pile at a time.

BUCKET DREDGE

Figure 10.7
Bucket–ladder dredge

The bucket dredge is appropriate for most types of soil but is best suited to medium sand, gravel, mud or clay, and loose rock fragments. Although the modern high-powered suction pumps and especially the cutting head dredge are proving more effective, the bucket dredge is probably still preferred where working in restricted space is required, for example, along quay walls and in dock systems. The method is also principally used for mining ores and precious metals.

The operation of the dredge consists of an endless chain of buckets continuously passing through the body of the barge to the working surface and is therefore only suitable in quiet waters. The buckets hold 50–1000 litre each but as a guide 50% of the contents is likely to be water, and output may range between 15 and 300 m^3. The barge is winched into position, can work to depths of 30 m, and discharges down a shoot into mud barges moored alongside. The chain and buckets are subjected to intense wear and the consequent maintenance costs are high.

DIPPER DREDGE

Figure 10.8 Dipper
dredge

The dipper dredge is a heavy-duty excavator operating with a similar action to the face shovel. The arm can turn through 180° and discharge to barges. The vessel is supported by two front piles, and a single pivoted rear pile to provide forward movement.

The main advantage is the powerful crowding action, so that the bucket can be forced into the underwater materials, which allows excavation in very compact and oversize materials such as rock, sandstone and coral, without the need for blasting. Depths up to 15 m may be worked and output of over 100 m^3/h is achievable with a 1½ yd^3 (1.14 m^3) bucket.

CLAMSHELL DREDGE

Figure 10.9 Clamshell
dredge

The clamshell dredge is little more than a pontoon-mounted grab crane and is used mainly for dredging in confined areas such as wharfs and breakwaters. The method is only suitable for use in loose soils and rocks.

GENERAL POINTS ON THE CHOICE OF DREDGING METHOD

(a) *Pure suction* is suitable in granular soil, such as medium sands with a particle diameter 0.2–0.6 mm which flow fairly easily.

(b) *Suction dragging* is suitable for use in most types of soil, but is most appropriate in naturally occurring fine sands and light loams, for example, 0.2 mm particle diameter and less.

(c) *Suction cutting* is best operated in clays, marl, gravel and soft rock. The clays in particular lend themselves well to transport to shore along pipes because of their ability to suspend in water.

Chapter 11   EARTH COMPACTING EQUIPMENT

Fill for the construction of dams, airports, roads, backfilling foundations and trenches is normally loose and bulked after the excavation process and must be compacted to prevent distortion, settlement and softening when in place. The attainable compaction, measured by the change in density, depends very much upon the type of material and moisture content, and whilst heavy rolling will significantly reduce the voids in granular soils, cohesive materials will consolidate only very slowly under a sustained pressure. In order to obtain satisfactory compaction in the vast range of soils to be met in practice, several different types of compacting plant have been developed, and are classified into those imparting static weight, impact, vibration or kneading.

STATIC WEIGHT ROLLERS

These rollers rely purely on self-weight to achieve compaction.

SMOOTH WHEEL ROLLERS

Figure 11.1 Smooth wheel roller (pulled type)

Single-axle Roller
The single-axle roller consists of a frame and smooth steel cylinder ballasted with sand or water to increase self-weight. As the roller is pulled forwards a wave of soil is pushed up (fig. 11.2) in the direction of movement and with successive passes over the ground surface the soil gradually compacts. The effect is more exaggerated with the pulled as compared to the self-propelled roller, and the latter is preferred in so far as the soil is disturbed less to produce a more even surface. However, for both methods the wave form will be larger as the weight of the roller is increased and the diameter decreased, the relative magnitude being related to composition of the soil. Such rollers may be used on most types of granular soils but are fairly inefficient when compared to similar machines fitted with a vibratory mechanism.

Figure 11.2 Rolling actions

PULLED ROLLER   SELF PROPELLED ROLLER

Three-wheeled Roller

This consists of a wide front roller and two narrower rear rollers and is predominantly used for compacting bituminous materials on road surfacing operations. Most three-wheel and also some tandem rollers are steered by the front drum with the drive provided through the rear rollers, usually set to overlap the track of the front roller by about 150 mm to prevent ridge formations. Differential gearing on the rear axle helps the machine to accommodate bends and curves without causing ruckling of the surface. In general, for all smooth wheel static rollers a thin layer near the surface is compacted very effectively, but unfortunately the soil below about 150 mm deep remains virtually untouched.

Figure 11.3 Three-wheel roller

Rubber-tyre Roller

It was observed that the constant pounding of traffic on a bituminous road surface caused grooving and further compaction of the road base and subgrade. On the basis of the evidence the rubber-tyre roller was developed. Today the method is mostly used for rolling base courses on roads and fill for large earthworks involving soil of a loamy texture. The machine is usually self-propelled and with ballasting, using either water, sand or pig iron, the weight may be increased by a factor of about 2.

Figure 11.4 Rubber-tyre roller

Rolling Method

In order to avoid ruckling of the surface when operating in curves and bends the wheels must have independent couplings combined with a swivel action to distribute the weight on each tyre evenly on undulating surfaces. Generally construction is with two axles comprising a total of seven tyres (three front and four rear) or nine tyres (four front and five rear), so arranged that the paths of the rear overlap those of the front to prevent ridges. A major advantage lies in the ability to control the ground contact pressure by:

(a) altering the weight of the machine;
(b) increasing the number of wheels;
(c) increasing the tyre width;
(d) changing the contact area of the tyre by altering the tyre pressure.

Manufacturers usually supply a table of ground contact pressures for given combinations of the above factors. Consideration of the bulb stresses under the tyres (fig. 11.5) indicates:

(a) the load on the tyre influences the depth to which compaction is possible;
(b) the tyre pressure and the tyre load are important in achieving high levels of compaction near the surface, but the tyre pressure becomes less important as the depth increases.

Figure 11.5 Bearing pressures under rubber tyres

Compacting Guidelines

(a) Loose lifts in sands and loamy soils according to Pohle of Aachen T.H., Germany, are:
    up to 300 mm layers (lifts) – tyre loads 1.5–1.7 tonnes
    up to 500 mm layers (lifts) – tyre loads 2.0–2.5 tonnes
    up to 700 mm layers (lifts) – tyre loads 4.0–4.5 tonnes
(b) Blacktop layers:
    up to 80 mm layers (lifts) – tyre loads 1.5 tonnes
    up to 130 mm layers (lifts) – tyre loads 2.5 tonnes
    up to 200 mm layers (lifts) – tyre loads 4.0 tonnes

When compacting earth, from four to eight passes are usually sufficient while for rolling blacktop, from four to six passes are recommended. Travelling first forwards and then backwards along the road at speeds up to 20 km/h produces the best results.

SHEEP'S-FOOT ROLLER

The sheep's-foot roller has projecting feet mounted on the surface, so-called because in the earlier models they resembled the shape of a sheep's foot. Roughly ten to twenty such feet are arranged around the rim at between 100 and 200 mm centres along the axis; several different shapes of foot are available to suit varying soil types. In general the method is preferred when attempting to compact highly cohesive materials. The small surface area of each foot transmits high pressure so kneading the soil particles together, and with repeated passes, the feet gradually 'climb' out of the fill. Practice has demonstrated that satisfactory compaction in a layer of fill equal to the length of a sheep's foot may be achieved.

Figure 11.6 Sheep's-foot roller

DYNAMIC COMPACTION

Such methods rely on either vibration or imparting kinetic energy.

VIBRATING ROLLERS

By setting the rim of a roller into oscillation, vibrations are transmitted to the soil particles and improve compaction. In comparison a much heavier static weight roller is necessary to achieve similar results. Today the sheep's-foot and smooth wheel rollers are constructed with a vibration unit, the frequency of which may be varied to render the method appropriate on most types of soil, including compacting broken rock.

Vibration Mechanism
Several systems are adopted by the various manufacturers, but most use the principle of the rotating eccentric weight, installed inside the roller drum (figs. 11.7a and 11.7b). Unlike the static weight machine the weight may not be increased by ballasting the drum but only by the addition of external weights which are usually hung from the support frame.

Figure 11.7(a) Roller vibration system

Figure 11.7(b)
Tandem roller (self-propelled)

Vibrating Plates

Two eccentric weights are placed either side of the centre of gravity of the plate and rotated out of phase in opposite directions. The resulting combination of forces impacts upon the soil surface to cause compaction, however, the horizontal components sometimes combine to act in the same direction when the vertical components are acting upwards allowing the operator to advance horizontally. By pivoting the whole block the movement may be set either for continuous forwards or backwards movement. The method is mainly used on fairly small patches and walked by the operator, it is more suited to granular materials.

Typically to achieve adequate compaction in cohesive materials a unit capable of delivering 1400 kg/m^2 is appropriate reducing to about 900 kg/m^2 for uniformly graded granular material. Approximately four to six passes may be necessary with layers up to 200 mm deep.

Figure 11.8 Vibrating plate compactor

Figure 11.9 Impact stamper

IMPACT PLATES

Kinetic energy is utilised by raising a heavy weight and allowing it to fall on to the surface. The common form is the hand-operated explosion stamper used in trenches and around small foundations and effective to approximately 500 mm depth.

OTHER TYPES OF COMPACTION EQUIPMENT

FREEFALL HAMMER

Again kinetic energy is used by dropping a heavy flat weight from the jib of a crane. In this way it is sometimes possible to compact layers of soil several metres below the surface.

Figure 11.10 Freefall
hammer

HIGH-SPEED COMPACTOR

Modern earthmoving projects such as motorways, earth dams and airports, involve large earthmoving machines and correspondingly rolling equipment must match the high production achieved by these methods. The introduction of the high-speed roller is proving very effective for this purpose. It comprises a pin-jointed frame for steering and is driven on four polygonal wheels, each of the front of which has from four to five polygon segments with the rear segments arranged to run between the furrows of the front wheels, so assuring complete coverage during the pass. The compacting action combines static weight, impact and kneading, and provides considerable advantages over other types of roller on massive areas involving cohesive and plastic soils, silts, gravel and hardcore. Adequate density results are possible with three to five passes with working speeds of 10–16 km/h.

Figure 11.11 High-
speed compactor

Table 11.1 Comparison of roller applications

Type	Self-weight (tonnes)	Ballasted weight (tonnes)	Drum or wheel diameter (mm)	Horse power, hp (kW)	Working speed	Frequency (Hz)	Loose lift (mm)	Remarks
Vibration	0.3–0.7	—	up to 1000	approx. 5 (3.7)	up to 3 km/h	up to 80	100–200	Trench reinstatement, small scale works and compacting bituminous surfaces
Vibration (including seated driver)	0.5–13.0	—	up to 1500	10–200 (7.5–150)	up to 10 km/h	up to 80	200–1000	Small hand-steered version similar to above, larger types with seated driver for large works, e.g. sub-bases and bearing courses on roads
Vibration	4.0–15.0	—	up to 2500	6–200 (4–150)	Towed	20–30	300–1000	Tractor-drawn suitable on large scale works, road subgrades, earth dams, heavy foundations, stone crushing
Static	4.0–15.0	up to 20.0	up to 2500	Towed	Towed	—	up to 300	
Static	8.0–14.0	up to 20.0	up to 2000	50–100 (37–75)	up to 10 km/h	—	up to 300	Mainly used for road surfacing operations
Rubber-tyred static	7.0–14.0	up to 30.0	—	100–200 (75–150)	up to 20 km/h	—	up to 700	Suitable for loamy soils on major construction, e.g. road bases, earth dams
Vibrating plate	0.1–1.0	—	—	2–10 (1.5–7.5)	up to 25 m/min	10–80	up to 700	Used on gravel and sand, very manoeuvrable, ideal for small scale work in awkward situations
Stamper	approx. 0.1	—	—	approx. 3 (2)	—	60–80 blows per min	up to 500	Very small scale works, compacting foundations, trenches, etc.

APPROXIMATE COMPACTION FACTORS

	In-place volume	*Bulked volume*	*Compacted volume*
Broken rock	1.0	1.5–2.0	1.3–1.4
Gravel	1.0	1.0–1.1	0.8–1.0
Clay	1.0	1.25–1.4	0.8–1.0
Sand	1.0	1.0–1.3	0.9–1.0
Common earth	1.0	1.1–1.3	0.8–0.9

Thus compacted volume = in-place volume × compaction factor, assuming of course that the material was compacted from a bulked condition.

SOIL CLASSIFICATION

Soil description	*Condition*	*Particle diameter (mm)*	*In place (bank) density (tonnes/m^3)*
	Dry		1.4–1.6
Sand	Damp	0.05–2.0	1.6–1.8
	Wet		1.8–1.9
	Dry		1.6–1.8
Silt and clay	Wet	0.001–0.05	1.8–2.1
Broken rock	—	50.0 up	1.5–1.7
Loam/	Dry		1.5–1.6
Common earth	Wet	Varies	1.6–1.7
Gravel	—	2.0–5.0	1.7–1.9

PART TWO

Notes

Chapter 12 SIMPLE LIFTING MECHANISMS

INTRODUCTION

The choice of lifting equipment is vast, ranging from simple ropes and pulleys, to large cranes and the trend is towards ever more sophisticated lifting devices. For example over 3000 tonnes capacity crawler cranes are currently in the design stage. However, even the large crane relies upon the principles of hoisting tackle, i.e. lifting rope, pulley blocks and a winch. This basic arrangement has been used successfully to design temporary works-lifting mechanisms for many tasks, including the erection of bridges, high-rise structures and industrial buildings. It is only during the past century that the modern crane has extended the application of such devices.

PRINCIPLES OF HOISTING TACKLE (ROPES AND PULLEYS)

Figure 12.1 shows a single line fixed pulley block, clearly P and W are only equal if there is no friction. When the block is free to rotate on an axis, the friction is a small component of the total forces and P and W can be assumed equal. Such a system, called 'single part' or 'fall', is commonly used for reeving on cranes.

The load-carrying capacity of the hoist tackle may be increased by using two, or more pulleys as in fig. 12.2. In the arrangement shown, if n is the number of ropes leading from the lower to the upper blocks, then neglecting friction, each rope supports W/n. Thus $P = W/n$. It is obvious, however, that considerably more rope must be wound in to raise W, compared with the

simple pulley

Figure 12.1 Single line pulley block

89

Figure 12.2 Blocks
and tackle

single line system. The velocity ratio of the system, defined as the ratio of the
distance moved by P to the distance moved by W is calculated as follows:

$$\text{Work done by } P = \text{work done by } W$$
$$PH = Wh$$

therefore $$\frac{H}{h} = \frac{W}{P} = \frac{W}{W/n} = n$$

In practice the pulley blocks at the top and bottom are mounted on a
single axis.

The reeving loads are governed by BS 1757:1964, which covers all the
possible reeving arrangements from single up to eight and above falls.

Figure 12.3 shows the methods of reeving for two, four, six and eight
falls. The rope follows the numbers on the pulleys beginning with the lead
line over the top block.

Figure 12.3 Reeving
methods

ROPES

6×37 Cable

Figure 12.4 Wire-
rope configuration

Cranes use wire ropes which comprise strands made up of individual wires,
of which there are several configurations as typified by the example shown
in fig. 12.4. The load carrying capacity of the rope depends upon its diameter
and the number of wires and is manufactured with a safety factor of approximately five.

The three common lays are ordinary lay. Lang's lay and non-rotating lay.

Figure 12.5 Ordinary right-hand lay

ORDINARY LAY

In the right-hand lay method the wire spirals to the left and the strands to the right (fig. 12.5). In the left-hand lay the arrangement is *vice versa* (fig. 12.6). These types of lay are useful as slings, but tend to wear quickly if operated as hoist ropes, since only the crown wires are in contact with the pulley.

Figure 12.6 Ordinary left-hand lay

LANG'S LAY

Lang's lay has both the wires and strands spiralling in the same direction (figs 12.7 and 12.8), but obviously both ends must be secured to prevent twisting. This lay has better wearing properties.

Figure 12.7 Right-hand Lang's lay

NON-ROTATING LAY

For most hoisting work with single part reeving the rope must avoid twisting. This is achieved by using a double rope construction, e.g. the inner rope can be in right-handed Lang's lay and the outer strands in left-handed ordinary lay. This method produces a non-rotating rope.

The size of the rope may be designed to suit the load requirements but for a medium size crane, e.g. 30 tonnes lifting capacity, an 18–20 mm diameter rope with up to six part reeving is usual.

Figure 12.8 Left-hand Lang's lay

WINCHES

All cranes and most lifting tasks with a block and tackle require a winch and rope drum to provide the lifting force. The winch may be powered by compressed air, electric motor, hydraulic motor or a diesel engine.

For portability around the construction site, the electric or compressed air winch is favoured. A typical arrangement of an electric winch is shown in fig. 12.9. The hoist speed of most winches may be varied to accommodate the line load. For example, with four-part reeving the described electric winch of 100 hp (74.6 kW) can handle 4000 kg at speeds up to 100 m/min, whereas at the full 20 000 kg load, the maximum operating speed is about 20 m/min.

Within these working ranges, both the lowering and raising speeds may be finely controlled, particularly on the modern independently powered electric or hydraulic winches. For the traditional diesel powered all-mechanical transmission winches, this is achieved by gear selection in conjunction with a brake and clutch arrangement. The hoisting speed may be further adjusted by altering the engine speed with the throttle.

Figure 12.9 Typical arrangement of an electric winch

Chapter 13   CRANES – SHEAR LEGS AND DERRICKS

INTRODUCTION

The applications of the simple block and tackle are somewhat restricted by the limitation of reach. Thus cranes of various forms have been designed to combine both horizontal and vertical movement, so providing a three-dimensional capability.

The earliest devices for use in civil engineering work were shear legs and later the derrick. Even today shear legs are widely used as a site-made temporary works crane, designed for special tasks peculiar to the particular duties called for on the project.

SHEAR LEGS

Where very heavy lifts are needed, then shear legs offer a very simple and inexpensive method. The equipment comprises a winch, and block and tackle suspended from a pin-jointed frame fabricated from steel tubing or RSJs, etc.,

Figure 13.1 Shear legs

Figure 13.2 Shear legs mounted on a pontoon

stabilised by guy ropes. Loads may only be raised and lowered, horizontal movement is not available. However, for river work, shear legs mounted upon a barge or pontoon can offer three-dimensional movement. Several bridges have been erected using this technique, but great skill in manoeuvring the vessel is required, and generally movement is only obtained by winching from anchorages located at convenient positions.

DESIGN PRINCIPLES

Figure 13.3 Forces acting on shear legs

Resolving horizontally

$$T \cos \alpha = R \cos \theta \qquad (1)$$

Resolving vertically

$$T \sin \alpha + W = R \sin \theta \qquad (2)$$

Thus T or R is calculated by solving equations (1) and (2). It is usual to apply a factor of safety of at least three to the forces in the members when selecting the actual material requirements.

GUYED DERRICK

The guyed derrick consists of a single boom and mast. The mast stands vertically and is guyed to anchorages in a similar fashion to shear legs. The arrangement allows both luffing (changing of radius) and slewing (turning),

Figure 13.4 Guyed derrick

but lifting is usually only attempted under a guy rope. It is possible to arrange for full 360° slewing, which makes for a very versatile crane and was at one time widely used where lifting facilities were required over a long period, and could justify the setting-up costs, e.g. the erection of steel-framed structures.

DERRICK

Figure 13.5 Derrick

The Scotch or stiff-leg derrick is more commonly chosen in preference to either the shear legs or guyed derrick, and is used for heavy lifting over long and high reaches. The derrick is extensively used for steelwork erection, especially in heavy plant construction, such as power stations and process plants, although with the continued development of the tower crane it is gradually losing its advantages.

CONSTRUCTION

The derrick consists of a vertical mast, usually made of steel plate, supported by two sloping fixed legs. The whole arrangement is seated upon a triangular frame of lattice construction, with the centre mast free to rotate on bearings at its top and bottom supports. The boom, attached to the base of the mast may be rotated through 270°, between the mast stays, and is capable of hoisting, slewing and luffing. The stays are relatively short compared to the length of the boom, and heavy ballasting is required at the base plates of the mast and stays. As a general rule the weight of kentledge at each ballasting point should be about four times the maximum lifting capacity of the crane. It is therefore essential that the unit is firmly supported on well prepared foundations. A common method of providing suitable foundations is to mount the derrick on bogies at the base apex points. The bogies themselves are supported on rails and sleepers, to provide the extra dimension of mobility.

The working height of the derrick may be increased by means of lattice towers, called gabbards, placed under the base apex points.

DESIGN PRINCIPLES

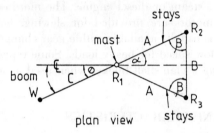

Figure 13.6 Forces acting on a derrick

(a) Taking moments about a line through R_2R_3

$$W \times (C \cos \theta + A \cos \alpha) = R_1 \times A \cos \alpha$$

$$R_1 = W \frac{(C \cos \theta + A \cos \alpha)}{A \cos \alpha} \qquad (3)$$

Therefore when θ increases to 90°, R_1 is equal to W.

(b) Taking moments about a line through R_1R_2

$$R_3 \times B \sin \beta = W \times C \sin (\alpha \mp \theta)$$

$$R_3 = W \times \frac{C \sin (\alpha \mp \theta)}{B \sin \beta} \qquad (4)$$

(c) Taking moments about a line through R_1R_3

$$R_2 \times B \sin \beta = W \times C \sin (\alpha \mp \theta)$$

$$R_2 = W \times \frac{C \sin (\alpha \mp \theta)}{B \sin \beta} \qquad (5)$$

It can be seen that as θ changes the reactions also change.

METHOD OF ERECTION

Complete erection, including commissioning, takes about 40 hours using three men and crane assistance. Dismantling can be achieved in about half the erection time. The following procedure is recommended:

(a) prepare the foundation bases and position the base frame;
(b) place the mast sole plate on the frame;
(c) erect the crane mast and temporarily guy to anchors;
(d) erect the stays in turn and secure with holding bolts;
(e) load on ballast;
(f) connect boom to mast;
(g) fit gears, winches, ropes, pulleys, etc.;
(h) raise the boom.

OPERATING THE DERRICK

The derrick is provided with two rope drums, one for derricking (i.e. luffing) and one for hoisting. Both are driven through gearing by an electric or hydraulic motor, steam or diesel engine. The more common form is electric, when separate motors are provided for slewing, hoisting and luffing. The hoisting facility is usually available with a gear change, the fast speed for light loads, and the slow speed for heavy loads. Some types of derrick have a third drum for opening a clamshell bucket when used as a grab.

DERRICK CRANE CHARACTERISTICS

The derrick is designed as a pin-jointed lattice frame structure to lift a certain load, called the design capacity, and is typically available with maximum load capacities up to 200 tonnes. Table 13.1 shows the corresponding maximum and minimum radius of operation for the more common sizes. It can be seen in fig. 13.7 that between the minimum possible radius (x) and maximum permitted radius (y) the load is limited to the design capacity load. For loads less than the design capacity the permitted operating radius (z) may be slightly increased. The corresponding values are shown in table 13.2. The working ranges for various lengths of boom are given in fig. 13.8.

Figure 13.7
Load–radius diagram for the derrick

Figure 13.8 Boom height–radius diagram for the derrick

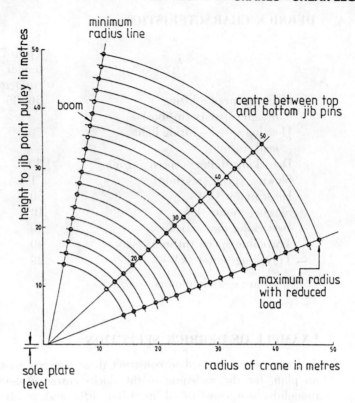

Table 13.1   Maximum operating radius at design capacity of derrick

Boom length (m)	Max. (y) radius (m)	Design capacity (tonnes)											
		3	5	7	10	15	20	25	30	35	40	55	
36	27	8	8.5	8.75	9	9	9.25	9.25	10	11	11	11	Min. operating
46	36	9	9	9.5	9.5	10	10.5	11	11	11	11	12	radius (x) (m)

Table 13.2   Permissible load at maximum operating radius of derrick (z)

Boom length (m)	Radius (z)(m)	Design capacity (tonnes)											
		3	5	7	10	15	20	25	30	35	40	55	
36	35	2.5	2.5	3	4.25	5	5.75	7.5	7.5	8.75	10	14	Load (tonnes)
46	42.5	2.5	2.5	3	4.25	5	5.75	7.5	7.5	8.75	10	14	at radius (z)(m)

DERRICK CHARACTERISTICS

	Up to 7 tonnes capacity	10 tonnes capacity and over
Hoisting speed – lifting design capacity (m/min)	30–35	10–15
Hoisting speed – lifting light load (m/min)	70	20–30
Derricking (luffing) speed (m/min)	30	12–15
Slewing speed (rev/min)	1	0.3
Hoist motor (kW)	50	50
Slewing motor (kW)	10	30
Derricking motor (kW)	40	50
Travelling speed (m/min)	40	10
Travelling motor (kW)	20	60

Note: All are electric motors.

EXAMPLE OF DERRICK SELECTION

A derrick crane is used to construct three concrete monoliths 20 m × 10 m on plan, for the widening of the docks entrance shown in fig. 13.9. The monoliths are constructed in 1.5 m lifts and a lift of reinforcing steel protrudes above the top of the shuttering. A maximum of four lifts of concrete is allowed above ground level as shown. The crane is to be used (i) for all materials handling including a 764 litre (1 yd^3) concrete skip, (ii) for excavating by grabbing action inside the monoliths with a 1000 litre (33 ft^3), 15° CECE rating, capacity grab, (iii) for lifting a complete cell of formwork weighing 3 tonnes.

Problem
Select a suitable size derrick crane.

Solution

(a) *Choosing the correct position*

Figure 13.9 shows that given the smallest available boom of 36 m and allowing two clear 1½ m lifts between the top of the reinforcement and the underside of the boom, the derrick can reach the four sides of the monolith when set up at its minimum operating radius of 8.5 m (position A on the section diagram), or alternatively at the crane's maximum radius (position C). This latter position however, is beyond the maximum permissible operating radius for the design capacity of the derrick. Clearly, therefore, the derrick can provide clearance at the maximum load permitted radius of 27 m (position B) and thus may be established in any position between points A and B. The top corners of each monolith, however, are outside the reach of the boom at all placings. Therefore the derrick should be rail mounted to cover all three monoliths.

Figure 13.9 Example layout of derrick crane

(b) *Choosing the appropriate crane capacity*
Possible loads are:
 (i) Shuttering – 3 tonnes.
 (ii) Concrete skip + concrete

$$
\begin{array}{ll}
\text{1 yd}^3 \text{ skip} & = 500 \text{ kg} \\
\text{1 yd}^3 \text{ concrete} & = \underline{1800 \text{ kg}} \\
& 2300 \text{ kg}
\end{array}
$$

(iii) Grab + contents

$$
\begin{array}{ll}
\text{1000 litre grab} & = 1350 \text{ kg (see table 14.4)} \\
\text{moist earth} & = \underline{1875 \text{ kg (heaped) (see table 14.2)}} \\
& 3225 \text{ kg}
\end{array}
$$

(iv) Hook block, etc. = 775 kg.
Max. possible load = 3225 + 775 = 4000 kg.
Include an extra 25% for surcharge, thus max. load = 4000 × 1.25 = 5000 kg.
A 5 tonnes capacity derrick is needed.

(c) *Ballast*
The ballast required on each foundation is approximately 4 × derrick capacity, i.e. 20 tonnes.

Chapter 14   CRAWLER-MOUNTED CRANE

INTRODUCTION

On many construction sites a crane is needed to lift small to medium loads, such as concrete skips, reinforcement, formwork, etc. Flexibility of movement around the site is at a premium, thus where the work is spread over a wide area beyond the reach of tower cranes and derricks, a crane capable of operating on unprepared surfaces is demanded. Rubber-tyred mobile cranes are excellent for lifting on level firm surfaces, but on many sites the ground conditions are so bad that these models would become bogged down and unable to work easily and efficiently. In conditions such as these the crawler-mounted crane is the most advantageous model to use. This is because the weight of the crane is spread over a large bearing area under wide and long tracks.

The crawler crane has the further advantage that conversion from crane to grabbing crane or dragline for excavation purposes is readily achieved.

CRANE CONSTRUCTION

The crane is built in three sections; the base frame, superstructure and boom, the whole unit being powered by a diesel engine.

Figure 14.1 Crawler-mounted strut-boom crane

THE BASE FRAME

The base frame is made from a welded steel channel to which the two machine axles are attached, and supports the weight of the engine, gearing and winches, controls, cab, boom and counterweight.

THE SUPERSTRUCTURE

The superstructure consists of a revolving frame sitting on a large turntable mounted on the base frame to give 360° slewing. The engine, gears, winches and counterweight are all mounted on this part of the unit.

The machine relies on a mechanical or hydraulic transmission system and two rope drums with independent brakes and clutches facilitate gravity lowering of the load. The front drum near the boom is the hoist winch and serves to raise and lower the crane hook and load, while the rear drum is used for luffing the boom.

PLAN VIEW

Figure 14.2 Gearing and winching layout (mechanical-type transmission)

THE BOOM

The basic boom is assembled in two sections each constructed from four high tensile rectangular hollow sections braced with round tube lacing members. The sections are pin-connected together and the top section incorporates a head sheave (fig. 14.3) for light loads of up to 10 tonnes or a hammerhead boom point (fig 14.4) for heavy lifts. Intermediate sections may be inserted to extend the length of the boom.

Figure 14.3 Simple head sheave

Figure 14.4 Hammerhead boom point

HOISTING TACKLE

The hoist rope between the head sheave and hook block may be arranged with one, two, three, four or more falls to accommodate the load being raised as described in chapter 12. A single part tackle naturally permits fast hoisting, but the weight of the load must be kept within the permissible tensile strength of the cable.

FLY-JIB

The straight boom suitably extended is ideal for general lifting duties but

Figure 14.5 Fly-jib

because of the inclined angle, obstructions often restrict load positioning. To overcome this problem a fly-jib may be attached to the boom point as shown in fig. 14.5.

The fly-jib is of similar construction to the main boom and is available in various lengths depending on the duties and capacity of the crane mounted in line with the main boom to act as a simple extension or more customarily at 30° offset to provide increased operating radius. If the crane is supplied with a third drum it is feasible to use this as the hoist winch with the fly-jib, leaving the main tackle on the boom for heavy lifts. Fly-jibs are designed for load lifting purposes only and are not suitable for grabbing crane or dragline operations.

WORKING LOADS

Figure 14.6 Forces acting on the strut-boom crane

It is preferable to operate the crane on a flat well-prepared surface. The tipping load may be determined by using the following formula.

$$L = \frac{W(p + y) - Zu}{x} - T$$

102

where

L = tipping load of the crane;

W = weight of the machine without the boom (but including counter-weights);

Z = weight of boom;

T = weight of head sheave;

R = radius to load from centre of slewing ring;

y = fulcrum distance;

p = centre of gravity of machine without boom to centre line of slewing ring;

v = connection point of boom to centre line of slewing ring;

x = $R - y$;

$$u = \frac{(R - v)}{2} + v - y.$$

Therefore safe working load $(P) = L -$ margin for safety.

Figure 14.7
Load–radius diagram for
the strut-boom crane

The shape of the load–radius diagram is shown in fig. 14.7. The weight of the hook block, together with any slings, etc., should be included when selecting a crane of suitable lifting capacity. If a fly-jib is attached but not in use then the safe working load should be reduced in accordance with the manufacturer's recommendations. Approximate load reductions are:

Fly-jib length (m)	Load reduction (kg)
6	850
9	1000
12	1200
18	1500
24	2100
30	2600
36	3000

Crane Radius Diagram

The precise range of boom lengths and the corresponding reaches and radii vary with individual manufacturers, but the dimensions shown in fig. 14.8 are typical of the makes of machine that are available.

CRANE CAPACITIES

The lifting ability at a given radius varies slightly with the particular crane manufacturer, with capacity being designated in terms of the maximum load

Figure 14.8a Boom height–radius diagram for strut-boom crawler crane: (i) Crane supported on tracks only: (ii) Crane on tracks and ringer system

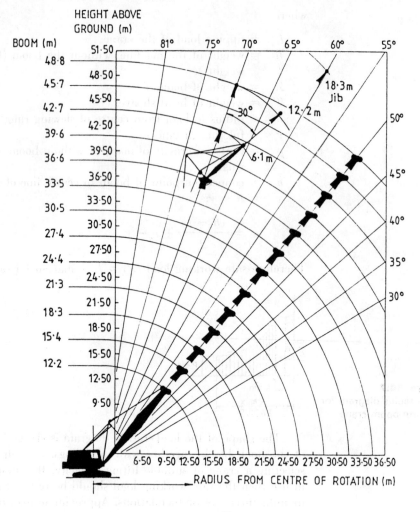

(i)

at the minimum operating radius. The most popular sizes are in the range 15 to 120 tonnes but cranes with capacities of 3000 tonnes or so are available. These larger units often have a ringer configuration (fig. 14.9), which as a rough rule-of-thumb doubles the SWL compared to the conventional tracked mounted version.

TRANSPORT AND ASSEMBLY

The crane is not intended to be self-transporting and will travel at little more than walking speed, and must be moved from depot to site on a low-loader truck. This choice is therefore not a practical option when lifting capacity is required for only a short period of hours or even a few days as the time taken to load, transport, unload and prepare for work may require a full day or more.

Figure 14.8b

(ii)

CONVERSION OF STRUT-BOOM CRANE TO GRABBING CRANE OR DRAGLINE (SEE CHAPTER 2)

Conversion to a grabbing crane or dragline can be readily obtained by changing the size of the hoist rope and hoist drum. The derricking drum is retained and an additional drum (i.e. third drum) included to control the bucket. Larger and wider tracks are often fitted to provide greater stability.

WORKING RANGES OF THE DRAGLINE AND GRAB CRANE

The dragline, like the crane, is usually designated with safe working loads

Figure 14.9 Ringer
configuration

based on a fixed percentage of the tipping load, depending on the country of operation, with the grabbing crane (and magnet cranes) normally being 80% of crane values.

The boom size may be varied to suit the required operating radius and lifting height as shown in fig. 14.10. As a rule-of-thumb the dragline is capable of digging to a depth below its tracks of roughly from one-third to half the length of the boom. Furthermore, the throw of the bucket beyond the radius of the boom may be of similar length, depending upon the skill of the operator.

The safe working loads of cranes shown in table 14.1 can be used to calculate the dragline working range, but should be reduced by 20% for grab cranes. Also, in practice, the working load may be further restricted in order to maintain an adequate factor of safety on the roping system, which will be in single fall reeving. The weight of the bucket must be included in the weight of the load to be handled, as recommended in tables 14.2, 14.3, 14.4 and fig. 14.11.

Figure 14.10 Working ranges of the dragline and grabbing crane

GRABBING CRANE

DRAGLINE

Table 14.1 Guidelines to safe working loads of crawler-mounted strut-boom cranes.

(a) Main Boom Rating: 15 Tonnes Crane

Radius (m)	Boom (m)			
	10.7	15.2	21.3	27.4
3.0	15.00			
4.5	8.6			
6.0	5.6	5.5	5.4	
7.5	4.2	4.1	3.9	3.7
9.0	3.2	3.1	3.0	2.9
10.5	2.7	2.5	2.5	2.4
12.0		2.1	2.0	1.9
15.0		1.5	1.4	1.4
18.0			1.0	0.9
21.0				0.8
24.0				0.7

(b) *Main Boom Rating: 20 Tonnes Crane*

Radius (m)	Boom (m)			
	10.7	*15.2*	*21.3*	*30.5*
3.0	20.3			
4.5	10.6			
6.0	6.6	6.4	6.4	
7.5	4.9	4.7	4.6	4.5
9.0	3.9	3.7	3.6	3.4
10.5	3.3	3.0	2.9	2.8
12.0		2.5	2.9	2.2
15.0		1.8	1.7	1.5
18.0			1.2	1.1
21.0				0.8
24.0				0.5

(c) *Main Boom Rating: 35 Tonnes Crane*

Radius (m)	Boom (m)					
	12.2	*15.2*	*18.3*	*21.3*	*24.4*	*27.4*
3.5	35.0					
6.0	17.0	16.9	16.8	16.7	16.2	
8.0	11.4	11.2	11.1	11.1	10.9	10.8
10.0	8.4	8.2	8.1	8.1	7.9	7.8
12.0	6.5	6.4	6.2	6.2	6.0	5.9
14.0	5.4	5.1	5.0	4.9	4.7	4.6
16.0				4.0	3.8	3.7
18.0				3.3	3.1	3.0
20.0				2.7	2.6	2.5
22.0					2.1	2.0
24.0						1.7
26.0						1.4

(d) *Main Boom Rating: 67 Tonnes Crane*

Radius (m)	Boom (m)					
	15.2	*21.3*	*27.4*	*33.5*	*39.6*	*45.7*
3.5	67.0					
6.0	33.0	31.9	30.0			
8.0	21.5	20.9	20.3	19.4	17.5	
10.0	15.5	15.1	14.6	14.2	13.9	12.3
12.0	12.2	11.8	11.5	11.2	11.1	10.4
16.0	8.1	7.5	7.3	6.9	6.7	6.3
20.0		5.3	4.8	4.6	4.3	4.1
24.0			3.8	3.4	3.1	3.0
28.0				2.7	2.2	2.1
32.0					1.5	1.4
36.0					0.9	0.8

(e) *Main Boom Rating: 91 Tonnes Crane*

Radius (m)	Boom (m)			
	18.3	27.4	36.6	45.7
4.5	91.0			
6.0	82.8	63.8		
8.0	42.9	42.1	41.1	
10.0	32.3	31.5	30.6	30.0
12.0	25.4	24.6	23.7	23.1
16.0	17.1	16.3	15.4	14.7
20.0		11.7	10.8	10.1
24.0		8.8	7.8	7.2
28.0			5.8	5.2
32.0			4.3	3.7
36.0				2.6
42.0				1.3

(f) *Main Boom Rating: 110 Tonnes Crane*

Radius (m)	Boom (m)					
	18.3	27.4	36.6	45.7	57.9	71.6
4.5	110.0					
6.0	77.6	22.0				
8.0	59.9	59.3	51.6			
10.0	46.8	46.2	45.7	35.2		
12.0	37.1	37.5	37.1	35.1	21.4	
16.0		25.4	24.9	24.7	20.1	12.8
20.0		18.9	18.5	18.3	17.6	12.3
24.0		14.9	14.4	14.2	13.6	11.8
28.0			11.7	11.5	10.9	10.2
32.0			9.7	9.5	8.8	8.2
36.0				8.0	7.4	6.7
42.0				6.3	5.7	4.8
48.0					4.4	3.7
52.0					3.8	3.1
58.0						2.2

(g) *Main Boom Rating: 217 Tonnes Crane*

Radius (m)	Boom (m)					
	18.0	27.0	36.0	45.0	57.0	63.0
4.5	217.8					
6.0	180.0					
8.0	122.5	122.0				
10.0	87.0	86.6	86.3			
12.0	67.1	66.7	66.3	65.7		
16.0	45.4	44.9	44.5	44.0	43.0	42.8
20.0		33.3	32.9	32.4	31.6	31.2
24.0		26.0	25.7	25.1	24.3	23.9
28.0			20.7	20.2	19.4	19.0
32.0			17.1	16.6	15.8	15.4
42.0					10.9	9.5
48.0					7.6	7.2
56.0						5.0

(h)   Main Boom Rating: 270 Tonnes Crane

Radius (m)	Boom (m)							
	21.34	27.43	36.58	45.72	51.82	57.91	64.01	73.15
5.6	270.0							
7.0	212.0							
9.0	148.4	148.2						
12.0	94.2	94.0	93.5	93.2				
16.0	62.3	62.00	61.4	61.0	60.5	60.2	59.9	59.4
20.0	45.7	45.4	44.8	44.5	44.1	43.8	43.4	42.9
30.0			25.4	25.0	24.6	24.3	24.0	23.5
40.0				16.1	15.6	15.2	14.9	14.4
45.0					10.7	10.3	9.8	9.2
50.0							6.8	6.1
55.0								5.0

Crawler crane.

(i)   Main Boom Rating: 550 Tonnes Crane

Radius (m)	Boom (m)												
	22.86	30.48	38.1	45.72	53.34	60.96	58.58	76.2	83.82	91.44	99.06	106.68	114.3
5.4	544												
7.0	442	440											
8.8	362	359	357	342									
12.2	229	228	228	225	230	230	227	224					
21.3	100	99	99	97	102	102	98	98	98	98	97	96	94
30.5		60	60	57	60	61	60	59	58	58	57	56	54
39.6				38	43	42	41	40	41	39	38	37	35
48.8					32	31	30	29	28	27	26	25	24
57.9						23	22	21	20	19	18	17	17
67.1							17	16	15	15	14	13	11
77.7									11	10	9	8	6

Crawler crane.

(j)   Main Boom Rating: 680 Tonnes Crane with 18 m dia. ringer

Radius (m)	Boom (m)						
	42.7	54.9	67.1	79.2	91.4	103.6	121.9
21.3	680	676	503	391			
30.5	370	381	379	353	273	217	149
39.6	232	261	258	256	247	197	134
48.8			193	190	188	176	118
57.9			147	149	147	144	102
67.1			107	118	115	116	88
77.7				85	92	92	57
82.3					81	85	40
86.9					70	75	—
91.4					61	55	—

Crawler crane.

(k) Main Boom Rating: 910 Tonnes Crane

Radius (m)	Boom (m)							
	47.2	62.5	77.7	92.9	108.2	123.4	138.6	153.9
13.7	907							
21.3	500	495	492	476	415			
30.5	295	290	287	284	281	252	224	161
39.6	205	200	196	193	190	185	170	128
48.8	153	148	144	141	138	133	129	99
57.9		115	111	108	105	98	96	75
67.1			88	85	82	77	73	53
77.7			71	65	62	57	53	43
82.3				59	56	50	46	41
86.9				52	49	44	40	37
91.4				48	44	35	35	32
93.5					38	32	29	25
103.6					33	27	26	20

Crawler crane.

Table 14.2 Approximate densities of soils

Material	Density, lb/yd³ (kg/m³)
Earth – moist	2500 (1490)
Sand – dry	2700 (1600)
Sand – wet	3300 (1960)
Gravel	2900 (1720)
Loose stone	2700 (1600)
Clay – wet	3000 (1780)
Coal	1350 (800)

Table 14.3 Dragline bucket data

Capacity	(yd³)	5	4½	4	3½	3	2½	2	1¾	1½	1¼
	(m³)	3.82	3.44	3.06	2.670	2.29	1.91	1.53	1.35	1.15	0.96
Weight	(lb)	9200	7700	7000	6400	5700	4700	4250	3300	2900	2300
empty	(kg)	4175	3495	3175	2905	2585	2130	1925	1495	1315	1040
Length	(m)	6.8	6.2	6.1	5.9	5.8	5.5	5.5	4.8	4.7	4.4

Table 14.4 Grabbing crane – medium-weight grabs

Capacity*	(ft³)	100/80	90/72	80/64	71/57	63/51	50/40	44/35
	(m³)	2.75/2.25	2.5/2.0	2.25/1.75	2.0/1.6	1.75/1.50	1.4/1.1	1.25/1.00
Weight	(lb)	6550	5200	5100	3900	3850	3050	2950
empty	(kg)	2975	2400	2350	1800	1750	1375	1350
Length	(m)	4.4	4.2	4.0	3.8	3.8	3.5	3.4

* Capacities given are heaped with 15° (CECE rating).

Figure 14.11 Dragline and grabbing buckets

Heaped   15°(CECE Rating)

Dragline bucket   Clamshell bucket

GROUND PRESSURE UNDER THE TRACKS

Pressure is given in N/mm^2.

	Crane capacity	
	30 tonnes	80 tonnes
Crawlers, standard tracks	0.06	0.113
Short crawlers, wide tracks	0.06	0.098
Long crawlers, standard tracks	0.055	0.076
Long crawlers, wide tracks	0.047	0.067

The pressure from a human foot is $0.02\ N/mm^2$.

LIFT CRANE DATA

Hoisting speed (single fall line)	approx. 40–50 m/min
Derricking (max. to min. radius)	approx. 50–100 s
Slewing speed	approx. 2 rev/min
Travelling speed	$2\frac{1}{2}$–3 km/h
Max. gradients when travelling: loaded	1 in 16
no load	1 in 5

EXAMPLE OF CRAWLER CRANE SELECTION

A crawler crane is used to lift a load of 5 tonnes (including hook block and tackle) from a temporary road (which acts as a flat and well-founded surface), to the centre of a reinforced concrete tank, as shown in fig. 14.12. The crane must provide $\frac{1}{2}$ m gap between the underside of the boom and the edge of the tank. Select an appropriate crane to perform the task.

Solution

(a) The crane is operated from the temporary road and with the gap between boom and tank maintained at the minimum distance of $\frac{1}{2}$ m a crane with a 21.34 m long boom is required. The alternative smallest next size of 18.29 m will not provide the necessary reach of 14 m radius from the centre line of rotation of the machine to the lifting hook.

Reference to table 14.1 reveals that the smallest size of crane capable of lifting 5 tonnes at 14 m radius is the 35 tonnes capacity crane. However, this may only be achieved with an 18.29 m boom, which is too short. The next size of crane would therefore be required.

Figure 14.12 Example of crawler crane in use

A cheaper alternative is to attach a 6.1 m fly-jib to a 12.19 m boom on the 35 tonnes capacity crane, (the lifting capacity of the crane is equivalent to the 18.29 m boom). The sketch shows that in this config-uration the desired operating radius of 14 m is achieved with some loss of operating height (which is not critical in this case). Thus a load of 5 tonnes may be lifted. It will be noticed that there is sufficient clear height between boom and tank to raise the load clear of the tank rim when slewing the load.

(b) The crane may be used for lifting with the main boom and sheave, with the fly-jib attached and out of action. The safe working load of the crane therefore must be reduced from the values shown in table 14.1 by 850 kg. Thus the maximum 5 tonnes load at an operating radius of 14 m with a 12.19 m boom is reduced to 4.15 tonnes.

(c) The crane may also be re-roped and operated as a dragline or grab crane. (Note – Only from the main boom, not the fly-jib.)

  (i) *Grab crane* – the lift crane safe working load is reduced by 20% for grabbing or clamshell purposes.

   Small bucket – 1250 m^3 heaped capacity bucket plus dry sand = 1350 kg self-weight + (1600 kg/m^3 × 1.25 m^3) = 3350 kg.

   From table 14.1 for a 35 tonnes capacity crane, the permissible operating radius is about 14 m with a 24.40 m boom, i.e. permissible load is 4.7 × 0.8 = 3.76 tonnes, which is greater than actual load of 3.35 tonnes. (Note – from fig. 14.10 jib angle is at the upper limit thus ruling out the longer 27.4 m boom.)

113

Large bucket – 2750 m³ heaped capacity bucket plus dry sand = 2975 kg self-weight + (2.75 m³ × 1600 kg/m³) = 7375 kg. From table 14.1 for a 35 tonnes capacity crane the maximum operating radius is about 9 m with a 18.2 m boom, i.e. interpolating the permissible load 9.5 × 0.8 = 7.6 tonnes, which is greater than actual load.

Note – from fig. 14.10 jib angle is near the limit.

(ii) *Dragline* – the lift crane safe working loads are directly applicable for dragline selection.

Thus using dragline data from table 14.3 a 1¼ yd³ (0.96 m³) capacity bucket + dry sand weight = 1040 + (0.96 × 1600) = 2576 kg. From table 14.1 for a 35 tonnes capacity crane, the permissible operating radius with a boom of suitable length, say 21.34 m, is 2.7 tonnes at 20 m. Assuming the bucket is cast one-third of the distance of the boom length beyond the boom, the boom could be set at an operating radius between 12 and 20 m, to keep within the permitted working range given in fig. 14.10.

A final check should be included to ascertain if the rope strength is sufficient to carry the working load.

Chapter 15   SELF-PROPELLED CRANE ON RUBBER-TYRED WHEELS

INTRODUCTION

Since the end of the second world war considerable growth in the use of cranes on wheels has taken place. The reasons for this are many but undoubtedly an important factor has been the demand for specialist one-off lifting facilities, effectively facilitated by the greater mobility afforded to cranes with the development of the diesel engine, efficient gear boxes and more recently by the introduction of telescopic booms – all aided by the crane rental market, which has allowed higher utilisation of specialist equipment than could otherwise have been achieved by individual construction companies owning their own plant.

Self-propelled cranes divide roughly into two classifications: the strut-boom type, but with the crawler tracks replaced by rubber-tyred wheels, and the mobile telescopic-boom vehicles. Whereas the mobile version is capable of travelling at 30–40 km/h on public highways, the 'converted crawler' crane achieves perhaps 8–10 km/h. The market trend seems to be moving away from the latter and may in the near future cease to be manufactured.

STRUT-BOOM CRANE

The crane is built in three sections similar to the crawler crane and comprises the base frame, superstructure and boom, the whole unit being powered by a diesel engine.

The base frame consists of a welded steel chassis and power is transferred to the wheels via a king-post gear passing through the turntable to a differential gearbox on the drive axle. The other axle is used for steering. The chassis usually has two-wheel drive, but for use on bad ground some machines are available with four-wheel drive.

The boom and winching arrangements are similar to the crawler crane, with all-mechanical transmission on the older versions and hydraulic motors on newer models. The machine may also be used in the grabbing mode, with suitable rearrangement of the drum sizes and ropes.

Figure 15.1 Self-propelled strut-boom crane

USES AND OPERATION

The crane is operated on fairly hard ground such as temporary hardcore roads in stockyards, in scrapyards (where it is usually fitted with a magnet attachment), and for lifting and transporting relatively light loads over short distances of a few tens of metres.

The load is transferred to the ground directly through the large rubber tyres, but some cranes' wheels are made with restrictor rims. There are two large-diameter heavy gauge steel discs mounted one each side of the rubber-tyred wheels. In effect they act as tyre stabilisers by restricting the flexing of the tyre and so dampen down the tendency of the machine to bounce when lifting and transporting. The restrictors also have a lipped rim which considerably increases the bearing area under load, particularly when operating from hard surfaces such as concrete.

OUTRIGGERS

To increase the operating range of the crane, outriggers are incorporated into the base frame on the larger versions, as shown in fig. 15.2. They are housed in heavy steel compartments and are extended and retracted by a winding mechanism. The effective width of the supporting base is thereby extended and any differential level of the ground can be taken out by adjustment of the pads attached to the extremities. The whole unit, including wheels, is

Figure 15.2
Outriggers for improved
stability

Outriggers

raised clear of the ground when lifting, all the load being transferred through
to the outriggers. This arrangement is called the *blocked* position. It is essential that the ground is well prepared and a solid foundation is available
particularly wherever heavy lifts are undertaken. The crane is of course
stationary in this configuration and is only used in this way for maximum
lifting.

CRANE CAPACITY

The crane size is defined by the maximum load at the minimum operating
radius. Lifting capacities *free* on the wheels, i.e. not *blocked*, range from small
cranes of 5 tonnes up to 15 tonnes. Typical capacity–boom–radius data are
shown in table 15.1 for free-on-wheels and blocked-on-outriggers arrangements. It is apparent from these data that the load–radius diagram for
wheeled machines is similar in shape to that of crawler cranes (fig. 14.7), but
suitable only for light lifting duties compared to the range of crawler cranes.

CRANE CHARACTERISTICS

Hoisting speed (single fall line)	40–60 m/min
Derricking (max. to min. radius)	20–40 s
Slewing speed	3 rev/min
Travelling speed (unladen)	10 km/h
Max. gradients: max. load	1 in 16
no load	1 in 8

Crane capacity (free-on-wheels) (tonnes)	5	10	15
Engine power, kW (hp)	30 (40)	37 (50)	52 (70)
Machine self-weight (tonnes)	14	18	28

CANTILEVER-BOOM CRANE

COMPARISON OF STRUT- AND CANTILEVER-BOOMS

A strut-boom offers the advantage of high lifting capacity when placing heavy loads at a wide radius, but being pin-jointed to the crane fairly close to the ground restricts the ability to lift and travel with wide loads when height restrictions are in force, for example, stockyard work where access into buildings is required. A cantilever-boom, however, is pivoted at a much higher position on the superstructure and so provides greater clearance, thereby facilitating the handling of bulky loads as shown in fig. 15.4.

Figure 15.3 Self-propelled cantilever-boom crane

Figure 15.4 Comparison between strut- and cantilever-boom cranes

(a) strut-boom   (b) cantilever-boom

DESIGN PRINCIPLES OF THE CANTILEVER-BOOM

The forces in a strut-boom are compressive, the tension component of the load being transferred to the superstructure through the derricking ropes. In contrast a cantilever-boom has a wide base tapering to the sheave support and is pin-jointed to the superstructure at the underside its lower end and restrained by the derricking rope running over a sheave along the top side as shown in fig. 15.5. The top member lattice frame is thus in tension and the bottom member in compression.

Figure 15.5 Cantilever-boom principles

CRANE CHARACTERISTICS

The choice of cantilever crane is limited within the range 5–15 tonnes free-on-wheels capacity.

Figure 15.6 Boom radius–height diagram for the cantilever-boom crane

119

Six Tonnes Capacity Crane (example only)

Max. safe working load	free-	6 tonnes as 3.0 m radius on 9.14 m jib
Max. radius and safe working load	on-wheels	850 kg at 13 m radius on 15.2 m jib
On outriggers		Not available
Max. hoisting speed (single fall line)		40 m/min
Slewing speed		3 rev/min
Derricking (max. to min. radius)		25 s
Travelling speed (unladen)		7 km/h
Weight of crane		16 tonnes
Engine horse power		48 kW (66 hp)

Eleven Tonnes Capacity Crane

The lifting characteristics are given in table 15.1. The weight of the crane is about 20 tonnes and is designed only for lifting *free*-on-wheels.

Table 15.1 Guidelines to safe working loads of strut-boom cranes on rubber-tyred wheels

(a) 7 Tonnes Crane FREE ONLY

	Boom (m)				
Radius (m)	7.6 FREE	10.7 FREE	13.7 FREE	16.8 FREE	19.8 FREE
3.0	6.0				
3.5	5.3	5.0			
4.0	4.2	4.0	3.5		
5.0	3.2	2.8	2.7	2.3	
6.0	2.5	2.1	2.1	1.8	1.7
7.0	1.7	1.6	1.6	1.5	1.3
8.0	1.4	1.4	1.4	1.2	1.1
9.0		1.1	1.1	1.0	0.9
10.0		1.0	1.0	0.9	0.7
11.5			0.9	0.8	0.6
13.0			0.7	0.6	0.5
14.5				0.4	0.3

(b) 11 Tonnes Crane FREE ONLY

	Boom (m)		
Radius (m)	6.1 FREE	9.1 FREE	12.2 FREE
3.0	11.0	11.0	
3.5	9.2	9.0	
4.0	7.5	7.4	6.2
5.0	5.5	5.4	5.2
6.0	4.3	4.2	4.1
7.0		3.4	3.3
8.0		2.9	2.8
9.0		2.5	2.4
10.0			2.0
11.0			1.6

Cantilever crane.

(c) 12.5 Tonnes Crane

	Boom (m)				
Radius (m)	9.1 BLKD	12.2 BLKD	15.2 BLKD	18.3 BLKD	21.3 BLKD
3.0	12.5				
3.5	10.6	10.2			
4.0	9.4	8.9	8.5		
5.0	7.4	6.8	6.6	6.0	
6.0	5.9	5.0	4.7	4.5	4.3
7.0	4.3	4.0	3.8	3.6	3.5

(d) 12.5 Tonnes Crane

	Boom (m)			
Radius (m)	9.1 FREE	12.2 FREE	15.2 FREE	18.3 FREE
3.0	7.0			
3.5	6.3	5.8		
4.0	5.4	5.2	5.1	
5.0	4.4	4.0	3.9	3.8
6.0	3.3	3.1	3.0	2.9
7.0	2.7	2.6	2.5	2.3

Table 15.1 Guidelines to safe working loads of strut-boom cranes on rubber-tyred wheels (contd)

(c) 12.5 Tonnes Crane

Radius (m)	Boom (m)				
	9.1 BLKD	12.2 BLKD	15.2 BLKD	18.3 BLKD	21.3 BLKD
8.0	3.5	3.3	3.1	3.0	2.9
9.0	3.0	2.9	2.7	2.6	2.5
10.0		2.5	2.4	2.3	2.1
11.5			1.9	1.8	1.7
13.0			1.7	1.6	1.4
14.5					1.1

(d) 12.5 Tonnes Crane

Radius (m)	Boom (m)			
	9.1 FREE	12.2 FREE	15.2 FREE	18.3 FREE
8.0	2.2	2.1	2.0	1.9
9.0	1.8	1.7	1.7	1.6
10.0		1.5	1.4	1.3
11.5			1.2	1.1
13.0			1.0	0.9
14.5				0.7

(e) 15 Tonnes Crane

Radius (m)	Boom (m)						
	9.1 BLKD	12.2 BLKD	15.2 BLKD	18.3 BLKD	21.3 BLKD	24.4 BLKD	27.4 BLKD
3.0	15.0						
3.5	13.2	12.6					
4.0	12.0	11.6	11.6				
5.0	9.6	9.0	9.0	8.5	8.3		
6.0	7.3	7.2	7.2	7.1	7.1	7.2	
7.0	6.0	6.0	6.0	5.9	5.9	5.9	
8.0	4.9	4.9	4.9	4.8	4.8	4.8	
9.0	4.1	4.0	4.0	3.9	3.9	3.8	3.6
10.0		3.5	3.5	3.4	3.4	3.3	3.2
11.5		2.9	2.9	2.8	2.8	2.7	2.6
13.0			2.6	2.5	2.5	2.4	2.3
14.5				2.0	2.0	1.9	1.8
16.0				1.6	1.6	1.6	1.5
17.5					1.4	1.3	1.2
19.0						1.1	1.1
20.5							0.9

(f) 15 Tonnes Crane

Radius (m)	(Boom (m))					
	9.1 FREE	12.2 FREE	15.2 FREE	18.3 FREE	21.3 FREE	24.4 FREE
3.0	7.0					
3.5	6.3	6.1				
4.0	5.6	5.4	5.3			
5.0	4.4	4.3	4.2	4.1		
6.0	3.5	3.4	3.3	3.2	3.0	
7.0	2.8	2.7	2.6	2.5	2.4	2.3
8.0	2.3	2.2	2.1	2.0	1.9	1.8
9.0	2.0	1.9	1.8	1.7	1.6	1.5
10.0		1.7	1.6	1.5	1.4	1.3
11.5		1.4	1.3	1.2	1.1	1.0
13.0			1.1	1.0	0.9	0.8
14.5			0.9	0.8	0.7	0.6
16.0						0.4

(g) 32 Tonnes Crane

Radius (m)	Boom (m)						
	7.6 BLKD	12.2 BLKD	15.2 BLKD	19.8 BLKD	22.9 BLKD	27.4 BLKD	30.5 BLKD
3.0	32.0						
3.5	31.0						
4.0	28.0	26.0					
5.0	23.0	22.0	21.0	18.0			
6.0	19.0	18.0	18.0	17.0			
7.0	15.0	15.0	15.0	14.0	13.5		
8.0		12.0	11.5	11.3	11.2	10.0	8.8
10.0		8.0	8.0	7.8	7.8	7.7	7.4
12.0			5.8	5.8	5.8	5.8	5.7
14.0			4.8	4.7	4.6	4.5	4.5
16.0				3.9	3.8	3.7	3.7
18.0				3.3	3.2	3.2	3.1
20.0					2.8	2.7	2.7
24.0						2.0	1.9
28.0							1.5

(h) 32 Tonnes Crane FREE ONLY

Radius (m)	Boom (m)						
	7.6 FREE	12.2 FREE	15.2 FREE	19.8 FREE	22.9 FREE	27.4 FREE	30.5 FREE
3.0	15.5						
3.5	13.8						
4.0	12.3	12.1					
5.0	9.8	9.7	9.7	9.7			
6.0	7.9	7.9	7.8	7.7			
7.0	6.5	6.4	6.3	6.3	6.2	5.5	
8.0		5.3	5.3	5.2	5.2	5.1	5.0
10.0		4.0	3.9	3.9	3.8	3.7	3.7
12.0		3.2	3.1	3.0	3.0	2.9	2.8
14.0			2.5	2.4	2.4	2.3	2.2
16.0				2.0	1.9	1.8	1.8
18.0				1.7	1.6	1.5	1.4
20.0					1.3	1.2	1.2
24.0						0.9	0.8

Chapter 16   MOBILE TELESCOPIC-BOOM CRANE

Since the early 1960s telescopic booms have been steadily increasing their share of the crane market at the expense of the self-propelled strut-boom model which has hitherto been very economical and efficient for recurrent usage on specific tasks, e.g. stockyard work. The mobile telescopic version is suitable for similar tasks but is more versatile and is capable of travelling at 20–30 km/h quickly moving from site to site in a particular area. The telescoping action is very flexible and various one-off lifts can be easily accommodated.

Unfortunately, the crane is expensive and as yet has not entirely replaced the self-propelled type for the duties required on the small-to-medium size site. In the long term, however, it must be considered as a serious competitor, particularly the specific versions developed to cope with rough terrain work, when the duties now fulfilled by the crawler crane, such as placing concrete, etc., may also be threatened.

CRANE CONSTRUCTION

CHASSIS

The chassis consists of two welded steel rectangular hollow side beams connected at each end by similar hollow boxes. A diesel engine is mounted at the rear of the vehicle and drives a rigidly mounted rear axle through a torque converter and power-shift gear box (see chapter 2, Fixed-position Excavating Machines, for details).

On some vehicles the rear as well as the front wheels have independent steering as shown in fig. 16.1. Steering on the latest models is controlled by hydraulic cylinders attached between the chassis and wheel hubs.

Two- or four-wheel drive actions are available, four-wheel drive is virtually standard on models made for rough terrain duties on construction sites. The superstructure comprises the drive cab and controls, telescopic boom, hoist and counterweight. The whole unit sits on a turntable, mounted on the chassis and thus 360° slewing is available. An independent hydraulic motor is used to induce the slewing motion. The hydraulic pressure is produced from a pump located near the engine on the base frame and thus a special rotary coupling located at the centre of the slewing ring is required to deliver hydraulic fluid to both the hoist and slewing motors.

Figure 16.1 Self-propelled telescopic-boom crane

Figure 16.2 Arrangement of superstructure

THE BOOM

The boom is designed on the cantilever principle shown in fig. 16.3. Whereas the self-weight of a strut-type boom is carried by the derricking ropes, the cantilever version is designed to support its own weight in addition to the

Figure 16.3 Principles of the cantilever–telescopic boom

load, the overturning moment being balanced by the counterweight with a suitable factor of safety. The boom is pin-jointed at its base to the superstructure and derricking is controlled by one or two hydraulic rams.

The boom itself consists of three or four sections, one sliding within the other. The telescoping action is provided by a hydraulic ram and a multiplying chain arrangement, as shown in fig. 16.4.

This configuration causes each section to be telescoped simultaneously and so ensures that the boom is maintained in a tapered shape to match the bending moment caused by the load.

Figure 16.4 Section through a telescopic boom

Boom Design

The restrictions on the design of the telescopic boom are those imposed by the self-weight and weight of the hydraulic telescoping rams which reduce the payload compared to the strut-jib. However, developments in metal strengths are gradually permitting the use of larger booms, and so increasing the operating range of the crane. Most manufacturers adopt a trapezoidal boom cross section made from high tensile alloy steel plate and sometimes the final jib section is of the lattice construction to reduce the weight of the boom.

125

HOIST OPERATION

The hoist is operated by a hydraulic motor driven from the main hydraulic pump. The hoisting and lowering speed may be varied for precision control of the load.

STEERING

Operation of the crane is carried out from a single cab mounted on the super-structure. The steering may be arranged to suit the duties required as follows:

(a) As a general purpose mobile unit the crane is either two-wheel drive from the rear axle (fig. 16.5a) or alternatively four-wheel drive for improved traction (fig. 16.5b) and is steered by the front wheels. Maximum travelling speed is about 50 km/h.

(b) To improve manoeuvrability, particularly for factory use such as stacking and warehousing duties, some versions of the crane have four-wheel drive and four-wheel steer (fig. 16.5d) which enables crab steering for diagonal movements (fig. 16.5c). Maximum travelling speed is 30–35 km/h.

(c) For heavy duty work, models equipped with the facilities described under (b) have a strengthened chassis and are provided with large diameter tyres for improved grip and flotation, and have increased ground clearance. These are called 'rough terrain' cranes and range up to about 150 tonnes maximum capacity. They are specifically designed to cope with the bad ground conditions found on construction sites. The tendency has been to try to use these cranes as genuine self-travelling cranes on the public

Figure 16.5 Steering alternatives

highways to increase the utilisation factor. But because they are not genuine on–off highway cranes, and due to the heavy maintenance requirements, they have not yet proved popular enough to supplant the crawler crane for difficult site work or the all-terrain model for on–off highway duties. Maximum travelling speed is less than 30 km/h.

OUTRIGGERS

The operating configuration may be unblocked on the tyres, but the lifting capacity is greatly improved when used with outriggers. Those shown in fig. 16.1 can be hydraulically operated and independently controlled from the driver's cab. They can be quickly set to provide a wide and stable base.

FLY-JIB

The main boom may be provided with the optional facility of a fly-jib which when not in use is stored on the side or underneath the boom. For lifting duties the jib is swung into position by means of the hoist rope and the guy ropes are attached. It may be used in line with the main boom or offset up to about 25°. A fly-jib option is usually available only on the large lifting capacity cranes of 10 tonnes and more – lengths up to about 8 m are available. On the larger rough terrain versions of the crane the fly-jib may also be telescoped (fig. 16.7) to improve the crane's reach for speedy applications.

Figure 16.6 'Quick assembly' fly-jib types

guy rope

Side folding fly-jib

guy rope

Under folding fly-jib

Figure 16.7 Telescopic fly-jib

CRANE CHARACTERISTICS

The smaller versions (up to 10–12 tonnes capacity) tend to be called 'mobile cranes', selected for stockyard work, whilst the larger and more robust machines are likely to be labelled 'rough terrain' cranes, being more appro-

Figure 16.8
Load–radius diagram for
a telescopic-boom crane

priate for site work. The shape of the load–radius diagram for these cranes is shown in fig. 16.8 and the load–radius data for individual cranes are given in table 17.1.

EXAMPLES

Four to Ten Tonnes Capacity Cranes Operating at Minimum Radius (i.e Small Cranes)

Min. lifting radius	2.5 m
Engine size	100 hp (74 kW)
Machine weight	12–15 tonnes
Max. hoisting speed (single fall line)	10 m/min on older models, 50–60 m/min on the latest models
Derricking (max. to min.)	10 s
Slewing speed	2 rev/min
Travelling speed (unladen)	30 km/h
Turning radius	6 m
Road gradient: loaded	1 in 6
no load	1 in 4
Min. boom length	4–5 m
Max. boom length	6–8 m
Overall height	approx. 2.75 m
Overall width	approx. 2.5 m
Overall length	approx. 6–7 m

The lifting capacity at maximum radius with the boom fully extended operating at an angle of 30° with the horizontal is about 1 tonnes.

Popular Rough Terrain Cranes

Safe working load at minimum radius on outriggers	15–40 tonnes
Engine size	150–200 hp (112–150 kW)
Machine weight	20–40 tonnes
Max. hoisting speed (single fall line)	up to 120 m/min
Derricking (max. to min.)	25 s
Slewing speed	3 rev/min
Travelling speed	30 km/h
Telescoping: three-part jib	20 s
four-part jib	40 s
Turning radius	10 m
Road gradient: unladen	1 in 25
Boom length: four-part	20 m
three-part	14 m
Overall height	3 m
Overall width	2.5 m
Overall length	8 m

Note: The safe working load in the *blocked* position is more than halved when lifting *free*-on-wheels only.

It may be observed by reference to tables in chapter 14 that telescopic mobile cranes as currently manufactured have a significantly lower lifting range of performance than the comparable size of strut-boom crawler models routinely used in placing concrete, handling reinforcement and bulky form-work panels. Thus when conversion to an excavating machine is considered as just one of the options available with the crawler crane, it is readily apparent that the wheeled crane is not particularly attractive for semi-permanent use on the construction site. Although this does not detract from the advantages of mobility between sites for short-term hire, and the large and more robust rough terrain of about 18 to 30 tonnes capacity suits the needs of a section of the market demand for cranage.

Chapter 17   TRUCK-MOUNTED TELESCOPIC BOOM-CRANE

While self-propelled and mobile cranes are suitable for on-site applications in stockyards, warehousing, dockyards, etc., modern construction sites often require cranes to provide medium-to-heavy lifting capacity over high and wide reaches. For example, placing precast concrete floor decks in high-rise construction, mechanical equipment in power station boiler houses, placing bridge deck beams, etc. Frequently the vehicle is only required for a short period of perhaps hours and fast travel between sites then becomes of paramount importance for economic viability. The obvious solution to the problem was development of the conventional truck or lorry to support a lifting unit, the first model being produced about fifty years ago, since then they have gradually become more efficient and reliable, with improvements made to the diesel engine and transmissions and more recently by the introduction of the telescopic boom.

The great advantage of the truck-mounted telescopic version is that travel at normal lorry speeds on the public highways is achievable and when on site takes only a few minutes to prepare for the lifting operations. Such a facility considerably reduces the hire cost, thus making the crane very competitive indeed.

Figure 17.1 Truck-mounted telescopic boom-crane

Cranes with a telescopic boom are available from about 10 to 800 tonnes capacity. However, vehicles less than 120 tonnes or so are increasingly being labelled 'all-terrain', and only the upper end of the range are true truck-mounted cranes. The all-terrain version (fig. 17.2) generally has all-wheel drive, with more robust construction and greater stability better suited to the temporary road surfaces found on typical construction sites. The latest models also have crab steering capabilities and improved speeds for highway travel. The boom on the all-terrain machine is usually shorter than the equivalent truck-mounted version to reduce overhang in the travelling mode, the chassis being much shorter. Finally, all-terrain models of 180 tonnes capacity are

Figure 17.2 All-terrain crane

now appearing on the market and the dividing line between these and truck-mounted units will inevitably be pushed to higher capacities.

Beyond these sizes telescopic-boom technology is currently inadequate and the conventional lattice boom must be used.

CONSTRUCTION OF THE VEHICLE

CHASSIS

The basic carrier comprises a chassis constructed from two universal beams with integral outrigger boxes. The chassis supports the power units, transmission, cab, boom, counterweight and hoists.

The vehicle has the appearance of a conventional truck and is described in terms of the total number of wheels and drive wheels, for example the truck in fig. 17.1 is designated 6 × 2 wheel drive. The number of axles required depends upon the travelling weight of the truck as the legal restrictions specifying the permissible axle load vary depending upon the country, typically 8–12 tonnes.

Figure 17.3 Chassis arrangement

SUPERSTRUCTURE

The upper part of the vehicle is similar in construction to mobile telescopic models, except that the counterweight is usually placed at a lower position to improve travelling stability. The whole unit is seated on a turntable to provide 360° slewing. On the smaller vehicles the crane-operating cab and drive cab are combined and a single diesel engine powers all mechanical parts, whereas larger versions have separate cabs and power sources for travelling and lifting duties. In this latter situation the hydraulic pumps and motors are all located on the superstructure and the hydraulic rotary coupling at the centre of the slewing ring is eliminated. The operating controls are also positioned on the superstructure to give the driver a clear view of the load and provide at least 180° of vision without having to look back, as is necessary when operating from the driver's cab.

Figure 17.4
Arrangement of
superstructure

CRANE CHARACTERISTICS

Max. hoisting speed (single fall line)	approx. 70–150 m/min
Slewing speed	3 rev/min
Derricking (max. to min. radius)	up to 1 min
Telescoping: in	10 m/min
out	20 m/min
Travelling speed	up to 70 km/h

TRAVELLING ON THE HIGHWAY

It is essential that the boom is fully retracted and well secured in the horizontal position, as shown in figs. 17.1 and 17.2 to avoid danger of damage to bridge decks.

CRANE CAPACITY

Mobile rough-terrain, all-terrain and truck-mounted telescopic boom cranes are designed in accordance with the following criteria:

(a) All capacities are limited to a proportion of the tipping load.
(b) With a fly-jib attached and extended but not working, the load on the main boom should be reduced, e.g. if length of fly-jib is 6–10 m, working load reduction is about 380 kg.
(c) Boom sections should be extended equally.
(d) The crane may be operated blocked on outriggers or free-on-wheels, with a corresponding reduction in capacity for the latter.
(e) The weight of the hook block, together with any slings or other lifting tackle, must be included in the working loads to arrive at a suitable size of crane.

Load-Radius-Boom Data

The data given in table 17.1 are general and should only be used for approximate crane selection purposes as the capacity of similarly designated cranes varies from make to make.

Figure 17.5 Boom radius–height diagram for truck/all-terrain cranes

Table 17.1 Guidelines to safe working loads of mobile, rough, all-terrain and truck-mounted telescopic-jib cranes.
(Note: Boom extensions quoted are maximum permitted values at the stated radius.)

(a) 6 Tonnes Crane Mobile   *(b) 12 Tonnes Crane Mobile*

Radius (m)	Boom m)				Radius (m)	Boom (m)			
	6.6 BLKD	FREE	8.7 BLKD	FREE		7.92 BLKD	FREE	18.3 BLKD	FREE
2.5	6.0				3.0	12.0	5.25	10.0	
2.75	5.5	5.25			3.5	10.5	4.5	8.6	
3.0	5.0	4.5	4.85		4.0	9.1	4.0	7.4	4.0
3.5	4.4	3.75	4.25	3.75	5.0	6.5	2.7	5.5	2.7
4.0	3.9	3.2	3.75	3.2	6.0	4.0	1.8	4.3	1.8

Table 17.1 Guidelines to safe working loads of mobile, rough, all-terrain and truck-mounted telescopic-jib cranes (contd)

(a) 6 Tonnes Crane Mobile

Radius (m)	Boom (m)			
	6.6 BLKD	FREE	8.7 BLKD	FREE
4.5	3.4	2.6	3.25	2.6
5.0	3.0	2.2	2.8	2.2
5.5	2.75	1.9	2.6	1.9
6.0			2.3	1.5
6.5			2.1	1.35
7.0			1.9	1.2
7.5			1.8	1.1

(b) 12 Tonnes Crane Mobile

Radius (m)	Boom (m)			
	7.92 BLKD	FREE	18.3 BLKD	FREE
7.0		1.1	3.5	1.1
8.0		0.6	3.0	0.6
9.0			2.4	0.4
10.0			1.9	
11.5			1.5	
13.0			1.1	
14.5			0.75	

(c) 18 Tonnes Crane Rough-Terrain

Radius (m)	Boom (m)						
	7.3 BLKD	9.0 BLKD	11.0 BLKD	12.2 BLKD	14.6 BLKD	16.5 BLKD	18.3 BLKD
3.0	18.3	15.6	14.8	14.0			
3.5	14.8	14.8	14.2	13.7	12.9		
4.0	13.4	13.4	13.1	12.9	12.0		
4.5	11.9	11.9	11.9	11.9	11.0	10.1	9.7
6.0	9.1	9.1	9.1	9.1	9.1	8.6	8.0
7.5		5.9	5.9	5.9	5.9	5.8	5.8
9.0			4.2	4.2	4.2	4.2	4.2
10.5				3.3	3.3	3.3	3.3
12.0					2.5	2.5	2.5
13.5						1.8	1.8
15.0						1.6	1.6
16.5							1.3

(d) 22 Tonnes Crane All-Terrain

Radius (m)	Boom (m)							
	9.75 BLKD	FREE	13.4 BLKD	FREE	17.1 BLKD	FREE	24.4 BLKD	FREE
3.0	22.0	7.0	19.6	6.7				
3.5	18.6	5.8	16.9	5.2				
4.0	16.0	4.7	15.1	4.3				
4.5	13.5	3.5	13.4	3.4	12.0	3.2		
6.0	9.8	2.2	9.5	2.0	8.6	1.8	6.5	1.1
7.5	6.5	1.2	6.5	1.1	6.0	0.9	5.2	0.7
9.0			4.5	0.6	4.5	0.5	4.0	0.3
12.0			2.5		2.5		2.3	
15.0					1.6		1.4	
18.0							0.9	
21.0							0.7	

(e) 30 Tonnes Crane Rough-Terrain

Radius (m)	Boom (m)					
	9.75 BLKD	13.25 BLKD	16.76 BLKD	20.37 BLKD	22.73 BLKD	30.48 BLKD
3.0	30.0	17.5	16.4			
3.5	26.0	17.5	16.4			
4.5	20.0	17.5	15.0	12.0		
6.0	16.0	14.5	13.0	10.5	8.5	
7.5	14.0	12.5	11.5	9.2	7.2	5.0
9.0		10.3	10.0	8.2	6.2	4.5
12.0			5.5	5.6	4.7	3.5
15.0				3.5	3.6	2.7
18.0				2.2	2.3	2.3
21.0					1.3	2.0
24.0						1.5
27.0						0.9
30.0						

(f) 40 Tonnes Crane All-Terrain

Radius (m)	Boom (m)						
	9.75 BLKD	11.6 BLKD	13.4 BLKD	15.2 BLKD	18.9 BLKD	23.8 BLKD	30.8* BLKD
3.5	41.0	35.0	30.0	27.0			2.5
4.5	32.4	28.8	26.4	23.9			1.7
6.0	24.0	24.0	22.8	20.8	16.8		
7.5	15.4	15.4	15.4	15.4	14.5	11.2	7.8
9.0		10.6	10.6	10.6	10.6	10.0	6.3
12.0				5.8	5.8	5.8	4.3
15.0					3.8	3.8	3.2
18.0						2.5	2.1
21.0						1.7	1.6
24.0							1.1
27.0							0.7
30.0							

(g) 60 Tonnes Crane All-Terrain

Radius (m)	Boom (m)							
	10.4 BLKD	13.4 BLKD	17.1 BLKD	20.1 BLKD	23.5 BLKD	27.1 BLKD	29.9 BLKD	36.2* BLKD
3.0	59.0	49.0						
3.5	54.0	46.0	39.0	32.0				
4.5	45.0	42.0	35.0	30.0	26.0			
6.0	36.0	36.0	30.0	24.0	22.0	21.0	16.0	7.5
7.5	25.0	26.0	25.0	20.0	18.0	16.0	13.0	7.0
9.0	18.0	18.0	18.0	17.0	15.0	14.0	11.0	6.0
12.0			10.0	10.0	10.0	9.0	7.0	5.0
15.0			6.0	6.0	6.0	6.0	6.0	4.0

Table 17.1 (cont)   *(g) 60 Tonnes Crane All-Terrain*

Radius (m)	Boom (m)							
	10.4 BLKD	13.4 BLKD	17.1 BLKD	20.1 BLKD	23.5 BLKD	27.1 BLKD	29.9 BLKD	36.2* BLKD
18.0					4.0	4.0	4.0	3.5
21.0						3.0	3.0	2.8
24.0						2.0	2.0	2.2
27.0							1.2	1.8
30.0								
33.0								

* Manual extension from top section.

(h) 90 Tonnes Crane All-Terrain

Radius (m)	Boom (m)						
	13.4 BLKD	18.2 BLKD	23.2 BLKD	28.0 BLKD	32.9 BLKD	42.7* BLKD	52.4* BLKD
3.0	90.0						
3.5	80.0	53.0	48.0				
4.5	69.0	51.0	44.0				
6.0	52.0	42.0	37.0	33.0	29.0		
7.5	39.0	35.0	31.0	26.0	23.0		
9.0	28.0	28.0	27.0	22.0	19.0	19.0	
12.0		17.0	17.0	16.0	14.0	14.0	10.0
15.0		11.0	11.0	11.0	10.0	10.0	8.0
18.0		7.0	7.0	7.0	7.0	7.5	6.5
21.0			5.0	5.0	5.0	6.0	5.5
24.0				3.0	3.0	4.0	4.5
27.0					2.0	3.0	3.5
30.0						2.5	2.8
33.0						1.6	2.2
36.0						1.0	1.7
39.0						0.5	1.1
42.0							0.6

(i) 112 Tonnes Crane All-Terrain

Radius (m)	Boom (m)						
	13.9 BLKD	21.3 BLKD	28.6 BLKD	36.0 BLKD	43.0 BLKD	52.8* BLKD	62.6* BLKD
3.5	112.0	53.0					
4.5	89.0	52.0					
6.0	67.0	45.0	40.0				
7.5	53.0	40.0	34.0	31.0			
9.0	42.0	35.0	30.0	27.0	24.0		
12.0		26.0	23.0	20.0	18.0	16.0	
15.0		18.0	18.0	16.0	13.0	13.0	9.0
18.0		12.0	12.0	12.0	11.0	10.5	7.5
21.0			8.0	8.0	8.0	8.5	6.5
24.0			6.0	6.0	6.0	6.8	5.7

(i) 112 Tonnes Crane All-Terrain

Radius (m)	Boom (m)						
	13.9 BLKD	21.3 BLKD	28.6 BLKD	36.0 BLKD	43.0 BLKD	52.8* BLKD	62.6* BLKD
27.0				4.5	4.5	5.3	5.1
30.0				3.0	3.0	4.5	4.4
33.0				2.0	2.0	3.3	3.6
36.0					1.0	2.4	2.9
39.0						1.8	2.2
42.0						1.1	1.6
45.0						0.6	1.1

(j) 150 Tonnes Crane Truck-Mounted

Radius (m)	Boom (m)					
	15.24	21.34	27.43	33.53	39.62	45.72
4.0	150.0					
6.0	100.0	98.0				
8.0	73.9	73.3	72.7			
10.0	56.2	55.9	55.700	55.4	52.0	
14.0	33.9	33.7	33.3	33.1	32.9	
20.0		20.4	20.000	19.8	19.5	32.7
24.0			15.6	15.2	15.0	19.3
30.0				11.0	10.6	14.3
35.0					8.4	10.4
40.0						7.9
						6.5

(R) 350 Tonnes Crane Truck-Mounted

Radius (m)	Boom (m)					
	16.75	28.75	40.0	52.0	72*	86*
3	350.0					
4.5	251.0	150.0				
7	163.0	117.0	80.0			
9	125.0	99.0	69.0	45.0		
11	99.0	85.0	60.0	44.5	25.5	
13	79.0	74.0	52.5	41.0	25.4	
18		47.0	39.0	33.0	21.5	9.0
24		28.0	29.0	25.5	17.7	8.4
28			22.0	21.5	15.7	7.6
34			14.5	16.6	13.1	6.4
40				11.8	10.8	5.4
46				8.5	8.6	4.5
54					5.0	3.5
64						2.5

Note: This is the largest crane able to carry its telescopic boom in the road travelling configuration.
* Including fly-jib.

Table 17.1 (cont)
(l) 500 Tonnes Crane Truck-mounted

Radius (m)	Boom (m) 18.6	31.4	44.2	57.0	57.0 Fly-Jib (m) 24	48	72
3	500.0						
4	380.0						
6	288.0	190.0					
8	230.0	189.0	140.0				
10	177.0	158.0	137.5	100.0			
15	115.0	105.0	101.0	80.5			
18		82.0	87.0	71.5	45.0		
22		60.0	65.0	60.5	44.0		
25		49.0	53.0	53.0	41.5		
30			38.5	43.5	39.0	22.0	
36			26.5	31.5		22.0	
40			20.5	25.0		22.0	8.8
46				18.0		21.7	8.4
50				15.0		20.2	8.2
60							7.8
70							7.0

(m) 800 Tonnes Crane Truck-Mounted

Radius (m)	Boom (m) 18	40	62	29 Fly-Jib (m) 23	51	93	62 Fly-Jib (m) 23	51	72
4	800								
6	500								
7	420	300							
8	360	260							
9	320	235	175						
10	290	215	165						
14	200	160	130	120					
18		125	102	117			67		
22		105	82	104	65		64		
26		85	70	92	63		59	37	
30		62	60		59		55	35	
34		48	53		54	16		33	21
40			37		47	14		28	19
48			23		38	13		26	19
60			10			11			16
72						9			12
84						8			

EXAMPLE OF TRUCK-MOUNTED TELESCOPIC CRANE SELECTION

Select a truck-mounted crane as an alternative to the crawler crane for lifting 5 tonnes to the centre of the concrete tank described in the example on page 112.

Solution

(a) It is seen in fig. 17.6 that to place the load a boom length of at least 18.9 m is required. Table 17.1 for a 40 tonnes capacity crane (which is sufficient in the case of a crawler crane) is slightly undersize. Interpolation between the 12 m and 15 m radii given in the table indicates that at 14 m radius the crane will lift about 4.5 tonnes. The next available size crane in the tables is 60 tonnes capacity, interpolating this will lift approximately 7.3 tonnes at 14 m radius.

(b) If the load were very wide and deep, say, half the diameter of the tank and 5 m deep, then to clear the tank rim whilst slewing would require a 20.1 m boom.

Figure 17.6 Example of truck-mounted crane in use

Chapter 18   TRUCK-MOUNTED STRUT-BOOM CRANE

Before the development of the telescoping action, the strut-boom was the predominant type of truck-mounted crane. Today, however, these cranes are limited to very specialist duties for lifts of 400 tonnes and more and are currently available with capacities up to 1200 tonnes. Designs are in hand up to 2000 tonnes.

CRANE CONSTRUCTION

The carrier consists of a stiff chassis mounted upon two or more axles depending upon the travelling weight. The chassis incorporates a diesel engine, the transmission and outriggers. The engine is used to power the vehicle for transporting only. The lifting section of the crane is a conventional strut-boom crane superstructure without the crawler tracks, which sits on a turntable mounted on the chassis. It comprises the counterweight, hoist and derricking drums, slewing mechanism, boom and operating controls, and an independent diesel engine to power the crane parts which like most other modern machines use independent hydraulic motors to provide precision hoisting and lowering control.

BOOM ASSEMBLY

This type of crane is frequently selected because of its ability to provide very specific heavy lifting capacity, economically over a short period. To do this effectively, the crane should ideally be ready and prepared for use almost immediately upon arrival at the construction site. The strut-boom unfortunately, however, cannot travel in this ready form, and must be folded and securely held. Consequently much setting up time is needed upon arrival, for example 4 hours preparation time would not be untypical. For use with the basic length boom, i.e. top and bottom sections only, the equipment packs down neatly as shown in fig. 18.1. But extensions beyond this basic length require the sections to be laid end-to-end on an area of level ground before erection. The assembly then proceeds as shown in fig. 18.2.

These very large capacity cranes often require two carriers; one for the crane (i.e. drive cab, engine, turntable, counterweights, hoist, etc.) and

Figure 18.1 Truck-mounted strut-boom crane arranged for travelling on a highway

Figure 18.2 Assembly of strut-boom

STAGE 1

STAGE 2

STAGE 3

STAGE 4

Figure 18.3 Large truck-mounted crane (chassis not shown)

one to transport the boom and fly-jib. The mechanisms for slewing, hoisting and derricking of this size crane are usually driven by electric motors powered from the main diesel engine. Extra outriggers and kentledge are also necessary (fig. 18.3) and the setting up time is much increased. The size of the load to be placed, however, is usually so specified that this can be taken into account in the construction programme.

CRANE CHARACTERISTICS

Travelling speed	up to 75 km/h
Slewing speed: up to 500 tonnes	0.5–1.0 rev/min
500–1000 tonnes	0.1–0.3 rev/min
Derricking (min. to max.)	Varies, e.g. 3 min for 300 tonnes crane with 15 m boom
Max. hoisting speed (single fall line)	60–120 m/min
Optional features	Operation as a tower crane

Load–Radius–Boom Capacity

The weight of the snatchblock, slings and other handling devices are significant and must, of course, be added to the load. The crane may be operated with a fly-jib and can be used either blocked-on-outriggers or free-on-wheels. Table 18.1 show the lifting capacities of 450 and 800 tonnes capacity cranes, respectively.

It will be noticed that these machines have very high lifting ability and are particularly suited for heavy duties, such as in offshore work, boiler installation in power stations, etc. The number of such cranes in the country are few and lifts must be planned well in advance to ensure availability of a crane of suitable capacity.

Table 18.1 Guidelines to safe working loads of truck-mounted strut-boom cranes
(a) Main Boom Rating : 450 Tonnes

Radius (m)	Boom (m)						
	16 m	30 m	44 m	58 m	72 m	86 m	100 m
5	450						
7	350						
9	295	292					
12	210	205	200	192			
14	176	174	170	160	125		
18		131	127	124	100	77	54
24		95	92	90	75	59	31
30		75	72	69	61	47	24
40			51	47	44	38	18
56				27	24	20	10
68					16	14	4
80						7	1

(b) Main Boom Rating : 800–1200 Tonnes

Radius (m)	Boom (m)							
	23 m	35 m	47 m	59 m	71 m	83 m	95 m	113 m
5	800							
7	650							
9	529	526						
10	478	475(1200)	470					
12	401	397(1000)	393	363				
18	268	265(645)	261	258	205	184		
22	216	214(600)	210	207	176	165	139	93
26		178(504)	174	171	152	144	126	85
32		135(410)	131	128	126	116	108	75
40			94	91	89	86	84	62
56				54	51	48	46	41
64					40	37	34	30
76						26	23	18
84							17	12
88								10

Note: Figures in brackets for 35 m boom are with alternative counterweight.

Chapter 19 TOWER CRANES

On travels far and wide the number of tower cranes in use is quite striking, with the skyline of the major cities often cluttered by new high-rise constructions. The tower crane, however, is also very suitable for low-rise work concentrated within a limited area where access by crawler or other mobile cranes is restricted.

The main advantage of the tower crane is that the jib or boom is supported at the top of a tall tower which may be set at a sufficient height to clear any obstructions. This configuration allows the crane to stand very close to, or even in, the structure under construction. In this way a relatively short boom provides more reach in comparison with other types of crane. Thus for high-rise buildings the tower crane is often the cheapest form of device.

There are two versions, the horizontal-boom and the luffing-boom, both powered electrically with 350–415 V supply.

LUFFING-BOOM

This arrangement consists of a lattice-framed vertically standing mast mounted on a sturdy turntable. The boom is hinged near to the top of the

Figure 19.1 Luffing-boom tower crane (turntable at base of mast)

144

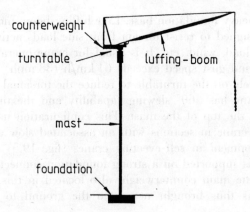

Figure 19.2 Luffing-boom tower crane, alternative kentledge at top of mast

counterweight

turntable

luffing-boom

mast

foundation

mast and the luffing line, attached to the extremity, is passed over a pulley at the top of the tower then down to the kentledge to counterbalance the weight of the boom. The kentledge is generally located on the turntable but some manufacturers prefer to have both the turntable and kentledge at the top of the mast as shown in fig. 19.2 so that all the moving parts are above the structure under construction. The whole unit is placed on a well-prepared foundation. The crane is electrically powered and electric motors are used to drive the hoist and luffing winches and for slewing. The main advantage of the luffing-boom tower crane is that the boom can be raised clear of nearby obstructions when slewing.

HORIZONTAL-BOOM

The crane comprises a vertical standing lattice-framed central mast, which supports a horizontal-boom in two parts, the larger section being used for lifting and carries a 'trolley' or 'saddle' travelling on guides along the length of the boom (fig. 19.4). Thus the radius is changed by moving the trolley and not by luffing the boom. On the opposite side of the mast a shorter boom supports a kentledge block and serves as a counterbalance. The resistance to overturning when lifting (and from wind pressures) is transferred through the

saddle or trolley

tie ropes

load

boom weight

counterweight

mast

kentledge

Figure 19.3 Horizontal or saddle-boom tower crane

145

Figure 19.4 Saddle or trolley for horizontal boom crane

2 FALLS

4 FALLS

tower to a heavy foundation base. Like luffing-boom models, the tower crane must be designed to resist torsion from side loads acting on the boom, e.g. a swinging load, wind, etc. It is usual for crane operation to be suspended when the wind gust speed exceeds 61 km/h (38 mph) and the boom is left to swing freely on the turntable to reduce the torsional effect.

The crane has 360° slewing capability and the turntable is commonly mounted at the top of the mast. This configuration usually involves transporting the crane in sections with an associated slow assembly on site, but rapid development in self-evecting cranes (fig. 19.5) is taking place. The whole unit is supported on a strong foundation connected to the base of the mast with the main counterweight also located in this position. The centre of gravity is thus brought nearer to the ground to improve the crane's stability.

Electric power provides the drive for the slewing and hoist motors.

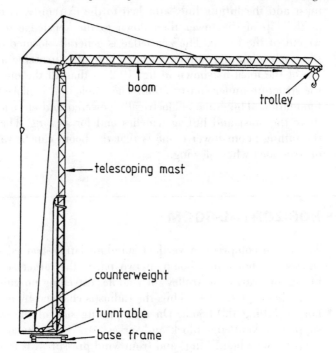

Figure 19.5 Self-erecting saddle-boom tower crane

LOAD–RADIUS DIAGRAM FOR THE LUFFING-BOOM CRANE

The tipping load of the luffing-boom is calculated from the following principles (excluding side loads).
Taking moments about the centre of rotation

$$Wr + J\frac{r}{2} = Cx$$

$$W = \frac{Cx}{r} - \frac{J}{2} \tag{1}$$

Figure 19.6 Forces acting on a luffing-boom tower crane

Figure 19.7 Load–radius diagram for the luffing-boom tower crane

Safe working load $(P) = W -$ margin for safety.
Equation (1) if plotted with varied r draws the curve shown in fig. 19.7.

LOAD-RADIUS DIAGRAM FOR THE SADDLE-BOOM CRANE

Taking moments about the centre of rotation

$$Wr + Jl = C_1x + C_2y$$

$$W = \frac{C_1x + C_2y - Jl}{r} \qquad (2)$$

Safe working load $(P) = W -$ margin for safety.
Equation (2) if plotted with varied r produces fig. 19.9, similar in shape to fig. 19.7.

For safety, the luffing-type crane operating radius of the boom is restricted to A min $\simeq 0.25\ A$ max. For the saddle-jib the minimum operating radius A min is limited to approximately $0.05–0.2\ A$ max. In general the load-radius diagram is further restricted as shown in fig. 19.9 for loads lifted near to the mast, and A m $\simeq 0.25–0.4\ A$ max. P min $\simeq 0.1–0.2\ P$. The saddle-boom crane is often quoted in terms of the load-moment capacity, which is calculated by multiplying the load by the radius. Usually the maximum load moment capacity occurs at A m and the minimum load moment at A max.

Tower cranes are now available to suit many situations and the data shown in tables 19.1, 19.2, 19.3, respectively for luffing-booms, saddle-booms and self- evecting models are typical of the free-standing capabilities with the crane operating with the longest possible boom.

Figure 19.8 Forces acting on a saddle-boom tower crane

Figure 19.9
Load–radius diagram for the saddle-boom tower crane

Table 19.1 Luffing-boom load-radius data

Max. radius	m	18	26	30	36	42	50	56	75
Lifting capacity	tonnes	0.75	1.1	0.6	1.0	1.4	1.8	2.0	5.6
Max. lifting capacity	tonnes	1.2	5	5	11	12	21	16	45
Radius	m	7	6	7	6	8.5	8.5	13	19
Height to jib pivot	m	18	27	29	35	46	52	58	76
Max. hook height	m	27	49	52	61	75	90	106	140
Track gauge	m		2.35	2.8	3.8–4.0	4.4–4.5	5.0	6.3	6.3–7.1	10.0
Power supply	kW	20	40	50	70	80	120	150	260

Table 19.2 Saddle-boom tower crane load-radius data

Max. jib radius	m	36	40	45	50	60	70	70	80	80
Capacity	tonnes	1.0	1.5	2.5	2.9	3.6	5.0	12.2	14.5	22.8
Max. capacity	tonnes	3.0	8.0	10.0	12.0	20.0	20.0	50.0	64.0	64.0
Radius	m	14.4	10.6	14.0	14.6	16.3	22.4	20.0	23.8	34.8
Max. load moment	mt	43	85	140	175	326	448	1000	1523	2238
Min. radius	m	7.0	4.0	3.0	3.0	3.0	3.9	3.9	4.0	4.0
Height under hook	m	30	38	44	47	66	80	90	140	105
Track gauge	m	2.8	4.5	4.5	4.5	8.0	8.0	10.0	15.0	15.0
Power supply	kW	40	40	60	100	150	150	170	200	200

Table 19.3 Self-erecting saddle-boom tower crane load-radius data

Max. jib radius	m	11	16	18	20	25	30	35	35	40	45	50
Capacity	tonnes	0.3	0.65	0.75	0.8	1.0	1.0	1.0	3.0	1.5	1.75	2.0
Max.capacity	tonnes	0.45	1.5	1.5	1.7	3.0	4.0	6.0	8.0	8.0	8.0	10.0
Radius	m	7.8	8.2	10.3	11.42	10.3	9.4	8.8	14.9	10.7	13.4	14.0
Max. load moment	mt	3.5	12.3	15.5	19.4	30.9	37.6	52.8	119.2	85.6	104.8	140.0
Min. radius	m	3	3	3	3	3	3	3	3	3	3	4
Track/wheel gauge	m	2.0	2.32	2.8	3.2	2.8	3.8	4.5	6.0	5.0	5.0	6.0
Power supply*	m	16	16	16	22	20	25	40	35	50	50	60
Max. hook height	kW	10.6	16.0	18.0	18	20	20.0	32.8	29.3	32.8	37.8	32.8

*Self-contained generator available on some models.
N.B. Available (1) static on outriggers; (2) travelling on rail track; (3) on crawlers.

TYPES OF TOWER CRANE (BOTH LUFFING- AND SADDLE-BOOM TYPES)

FREE-STANDING CRANE

The free-standing crane, stands self-supporting on a well prepared foundation. The mast is bolted to a strong steel cruciform base (fig. 19.10). Ballast consisting of blocks of concrete, iron or gravel, is placed on the base to provide a counterbalance.

Free-standing cranes are available up to 100 m clear distance between the hook and ground level. To increase the height further the crane must be tied to the structure under construction.

The free-standing arrangement is commonly adopted and includes both the saddle- and luffing-boom types. It is usually preferable to position the crane outside the building, as this makes for easier erection and dismantling and also avoids the costs of leaving out and making good parts of the structure.

With the fixed-position base the crane boom must be capable of reaching all parts of the job and so tends to be used for building construction and particularly high-rise construction where regular shape structures are involved as typified in fig. 19.11.

RAIL-MOUNTED CRANES

An alternative and popular foundation is the rail-mounted base. With this arrangement the crane can travel with the load on the hook on a specially prepared track, which may be straight or curved.

The base frame is made from sturdy welded steel plate. For straight track the frame is of simple construction mounted on four wheels as shown in fig. 19.13. For curved track (fig. 19.14) the frame is more sophisticated and the

Figure 19.10 Free-standing tower crane base layout

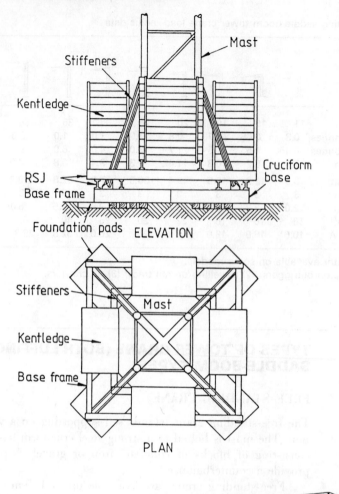

Figure 19.11 Free-standing tower crane applications

arms and wheels are pivoted to accommodate the bends in the track. Usually two wheels are provided under each support. The crane can be used to construct both high and long buildings and is able to travel to material stock-piles, concrete batching plants, etc., located in close proximity to the structure.

There are some disadvantage compared with the free-standing tower crane, as follows:

Figure 19.12 Rail-mounted tower crane

ELEVATION

PLAN

Figure 19.13 Tower crane base for straight rail track

Figure 19.14 Tower crane base for curved rail track

(a) Track is expensive to purchase and lay.
(b) The site must be firm and the crane can only be used on a level track, rails out of level by only a few millimetres render the mast unstable.
(c) The crane's travel motors can accommodate gradients only up to about 1 in 200.
(d) The track must be kept clear at all times, which may entail supervision, especially on a busy site, and so add to the costs.
(e) The track may occupy a valuable area of the construction site, which cannot be used for storage of materials.

Rail-mounted cranes are electrically powered with separate three-phase a.c. motors, requiring a high voltage supply of 350–415 V. The travel wheels

151

are driven through gears to enable the crane to move as fast as is safe with the particular load being carried. Sometimes the crane is provided with a special braking system to allow fine movement when precision positioning of the load.

CRAWLER-MOUNTED TOWER CRANES

There has been a limited demand for tower cranes with greater mobility than is provided by rails. For example, crawler-mounted tower cranes (fig. 19.15) have been successfully used to pour the concrete foundations and floors, and lift out the waste materials for the construction of housing and low-rise dwellings. For economic application, however, duties would need to include the placing of all materials plus positioning any units such as precast floors and walls. Such cranes, however, are relatively small – up to 1500 kg capacity at 40 m radius. The crane is brought to site on a low loader.

Figure 19.15 Crawler-mounted tower crane

TIED-IN TOWER CRANE

The free-standing tower crane supported on a ballasted base or rails can be used up to heights of about 100 m. Above this the crane must be tied in to the structure under construction at points recommended by the manufacturers of the crane, to coincide with suitable positions on the building, such as the floors of a high-rise dwelling. The tie frame is of special design usually in the form of a lattice frame which spreads out from the mast and is attached to the structure to resist forces from all horizontal directions. By careful selection of the fixing positions the operating height may be extended up to 200 m or more. The base may be ballasted as in fig. 19.10 but frequently, where the site conditions allow a concrete foundation block is cast into the ground as shown in fig. 19.17. The mast is then simply bolted down. Most of the overturning moment is taken by the structure rather than the foundation, resulting in a lighter and neater base than either rail or ballast types, with a consequent increase in working space on a congested site.

TRUCK-MOUNTED TOWER CRANE

The strut-boom truck-mounted crane of the lattice type may be assembled to operate as a luffing tower crane as an optional feature on the large cranes

Figure 19.16 Tied-in tower crane

ties to building

Figure 19.17 Base for a tied-in tower crane

bolts

100 mm

Reinforcement

Nuts

Anchor plate

Figure 19.18 Truck-mounted tower crane

of 100 tonnes capacity and greater. The fly-jib is replaced by a much longer luffing boom which operates with the safe working load at 75% of the tipping load. The crane is suitable for short duration one-off tasks where heavy loads need to be lifted into high wide structures. Lifting data are given in table 18.1.

JACK-KNIFE CRANE

The jack-knife crane is designed to work in extremely tight quarters, for example, work between two high buildings. A free-standing horizontal boom crane would have to free sail over the top of such structures and thus might pose an unnecessarily expensive solution to the problem. Capacities are available for all the common site duties and permit about 2 tonnes to be raised at 30 m radius.

Figure 19.19 Jack-knife tower crane

CLIMBING CRANE

Where external site space is at a premium and the shape of the building allows, a tower crane may be located inside the building. Frequently the lift shaft in high-rise structures serves as a convenient position to locate the mast. Generally the tied-in crane referred to earlier is used in this situation but where the client permits, a climbing crane may be cheaper, as the foundation and most of the mast is dispensed with. The horizontal and vertical thrusts must therefore be taken by the structure itself.

Initially the mast is mounted on a fixed base which is usually part of the foundations of the structure. The crane then proceeds to erect the building around itself up to at least the second floor. It is then secured by special collars at each floor level. Thereafter it climbs without support away from the original foundations base. The two principal methods of raising the mast are by winches or hydraulic jacks.

Figure 19.20
Climbing tower crane

Winch Raising

A line (F) is attached to the collar (H) which is firmly fixed to the structure. The line is passed under two pulleys (E) fixed to the base of the mast and over a third pulley (G) connected to the pulley (D). A second line (J) from a winch (A) mounted at the base of the mast is passed over the pulley (C) which is hung from the top of the mast section, it is led down and under pulley (D) and finally attached firmly to the support at (C). The winch winds in rope (J) and raises the mast, lifting it clear of the bottom collar. The collar is then firmly clamped and the lifting tackle is positioned ready for the next lift.

Figure 19.21 Raising a climbing tower crane by winching

Hydraulic Raising

The lifting procedure starts with the mast in position (A) supported on a climbing ladder (1), hung from a bearing plate (6). The collar (2) is firmly clamped to the crane mast, with the spring claws (5) engaged on rungs (4). A sliding collar (3) is linked to collar (2) by means of the hydraulic jack (7). To raise the mast, the claws on collar (3) are positioned to engage ladder rung (8). The jack is extended and thereby lifts the crane from position (A) to position (B). The jack is then withdrawn raising collar (3) in the process.

Dismantling requires the boom and mast to be taken apart in sections and lowered to the ground by winches.

155

Figure 19.22 Raising a climbing tower crane by hydraulic jacking

---

ERECTION OF TOWER CRANES

TURNTABLE AT THE TOP OF THE MAST

SUPERSTRUCTURE

BASE FRAME

PREPARATION OF MAST

MAST IN SECTIONS

ASSEMBLY OF BOOM AND COUNTER BOOM

COMPLETION OF ERECTION

Figure 19.23(a) Example of a tower crane method of assembly

Figure 19.23(b)
Erection of a tower
crane

1. BASE ERECTION

2. MAST ERECTION

3. COUNTERWEIGHT
   BOOM ERECTION

4. MAIN BOOM ERECTION

An example of a winch-assisted method of erection is shown in fig. 19.22(A). For some makes the erection procedure is lengthy and usually requires the assistance of a separate crane usually a truck-mounted telescoping boom-type mobile. All the parts are brought to site on trucks and the crane is put together in sections as follows:

Stage 1: The base is prepared and the first section of mast is lifted into position.

Stage 2: The climbing cage and turntable are next positioned over the mast.

Stage 3: The counterweight and arm are fixed in place. (Saddle-boom type only.)

Stage 4: The boom is attached and finally self-hoisted or lifted by crane into position.

The crane height is further increased either by introducing mast sections at the top of the mast as shown in fig. 19.24 or alternatively through the side as in fig. 19.25.

Figure 19.24
Extending the height of
a tower crane from the
top

Figure 19.25
Extending the height of
a tower crane through
the side

TURNTABLE AT THE BASE OF THE MAST

Cranes of both the luffing- and saddle-boom types can be made self-erecting
by designing the mast on a telescoping principle. The operating height,
however, is limited to the design size of the crane.

The whole unit usually arrives in the collapsed arrangement towed by an
articulated wagon (fig. 19.26) or even as a completely mobile truck-mounted
crane (fig. 19.18).

Using built-in winches on the crane the tower and jib are unfolded as
shown in fig. 19.27.

Figure 19.26
Transporting a tower
crane on the highway

Figure 19.27 Self-erecting tower crane sequences

CRANE ASSEMBLY TIME

None-self-erecting cranes require 1 day for small units and 2 days for large units with crane assistance. Without crane assistane, using only winches, up to 4 days with three men may be required. Self-erecting cranes require from $\frac{1}{2}$ to 1 day. Laying of a 10 m length of rail track takes from 1 to 2 days. The ratio of erection to dismantle time is about 60:40.

TOWER CRANE CHARACTERISTICS

				Crane size small	medium	large
Max. hoist speed (two part line):	raise	V_R	m/min	30.0	55–90	150
	lower	V_L	m/min	80	100	150
Slewing speed		U	rev/min	1.0	0.8	0.6
Trolley speed		V_s	m/min	40	30	2
Travelling speed (on rails)		V_T	m/min	40	30	0
Luffing		t_D	s	20	50	16
						90

EXAMPLE: CYCLE TIME FOR A LUFFING-BOOM TOWER CRANE (see figs 19.12 and 19.28)

1. Hook on load $\qquad t_{o_1} = 1$ min
2. Raise load from ground level $\qquad t_{R_1} = (h + h_s)/V_R$
3. Slewing $(\beta_1 + \beta_2)$ $\qquad t_{s_1} = [\beta_1 + \beta_2)/360] \times U$
   $\qquad (\beta_3)$ $\qquad t_{s_2} = [\beta_3)/360] \times U$
4. Travelling along S (on tracks) $\qquad t_{T_1} = S/V_T$
5. Derricking jib into position $\qquad t_{D_1}$
6. Lowering load (h_S) $\qquad t_{L_1} = h_S/V_L$
7. Unhook load $\qquad T_{O_2} = 1$ min
8. Raise hook (h_s) $\qquad t_{R_2} = h_S/V_R$
9. Slewing (β_3) $\qquad t_{S_3} = [(\beta_3)/360] \times U$
10. Derricking (as 5) $\qquad t_{D_2}$
11. Lowering hook $(h + h_S)$ $\qquad t_{L_2} = (h + h_S)/V_L$
12. Travelling along S $\qquad t_{T_s} = S/V_T$
13. Slewing $(\beta_1 + \beta_2)$ $\qquad t_{s_4} = [\beta_1 + \beta_2)/360] \times U$

The bar programme is shown in fig. 19.28.
The number of cycles per hour (L_0) is given by

$$L_0 = \frac{60}{\Sigma t_0 + \Sigma t}$$

Figure 19.28 Tower
crane cycle time

This is the theoretical output which in practice may be reduced by other factors such as weather stoppages, breakdowns, etc. These may add up to total losses of 30%. The experience and skill of the driver are also important, for example the efficiency factor may be 0.95, 0.85 or 0.75 for good, average and poor drivers, respectively. Thus calling the production losses factor e_1 and the driver efficiency factor e_2

$$\text{expected output} = \frac{60e_1e_2}{\Sigma t_0 + \Sigma t} \text{ cycles per hour}$$

Typically $\frac{1}{2}$ yd^3 skip of concrete can place at the rate of about 7 m^3/h and the corresponding rate for a 1 yd^3 skip is 12 m^3/h at heights up to 25 m.

TOWER CRANE EXERCISE (1)

A tower crane required to concrete a 100 m long retaining wall is positioned as shown in fig. 19.29 such that the maximum length of boom is 20 m. The unit is mounted upon rails which run parallel and 20 m distant from the retaining wall. The crane data are as follows:

Raising speed	30 m/min
Lowering speed	60 m/min
Slewing speed	1 rev/min
Travelling speed	40 m/min
Derricking speed	40 s

The operating efficiency of the crane is 85%. The raising and lowering height of the load from ground level is 20 m. The times to fill and empty the skip are:

Fill	30 s
Empty	30 s
Position over wall	30 s

Figure 19.29 Example of a tower crane in use

Problem

(a) How many skips per hour can the crane handle?

(b) How far along the track can the crane travel without influencing the output of the crane?

(c) Draw the crane's operating cycle diagram independent of the travelling time along the track.

TOWER CRANE EXERCISE (2)

Figure 19.30 shows a plan view of the present site layout for the construction of a rectangular high-rise building using a tower crane. Criticise this arrangement

Information   (i) 500 litre concrete batcher sited 40 m from tower crane.

               (ii) Load moment of tower crane at 40 m radius is 30 mt.

PRESENT METHOD

Figure 19.30 Example of a tower crane site layout

161

Solution

Present method

(a) Three entrances to the site (two not guarded) are required.
(b) A 500 litre batching plant is too large for the safe working load at the required radius.
(c) The mortar mixer is outside the reach of the crane.
(d) Materials are too spread over the site.
(e) Office and stores are too far from the site.

Proposed method

(a) Stores and office are located near the site entrance and exit.
(b) A through road now connects the entrances and exit.
(c) The mortar mixer is located within reach of the crane.
(d) The materials compound is arranged along crane track.
(e) The mixer is reduced to 375 litre to match the tower crane capacity.
(f) The third entrance is eliminated.

Chapter 20   GANTRY (PORTAL) CRANE

The gantry crane is usually associated with stockyard work, for example precast concrete manufacturing where lifting facilities are required to cover both the casting and storage areas, for which a portal structure is ideally suited. The portal legs are mounted upon rail track and the bridging beam is positioned to span the full width of the manufacturing/stockyard area. The structure is of lattice frame construction, with one pair of legs rigidly mounted to the bridging beam to give stability and designed to resist both horizontal and vertical forces, while the other pair of legs is of a more slender construction and carries only the vertical component of the load force. The portal beam supports an electrically powered hoist travelling on wheels along the bottom flange, generally arranged to cantilever some distance out from the legs with height altered by changing the number of sections. Models are available with capacities ranging from 5 to 30 tonnes. Purpose-built cranes are designed to suit specific needs.

Figure 20.1 Gantry or portal crane

TYPICAL DATA

Clear span	25–50 m
Cantilever, each side	5–15 m
Clear height below hook	5–10 m (may be increased for specific requirements)
Wheel base	10–15 m
Travel speed	30–40 m/min

USES

The working height is a limiting factor in comparison with, for example, tower cranes on high-rise construction, but occasionally where the high capital cost of the gantry crane can be justified on a single contract or over a series of similar contracts, adoption of this method of materials handling is very efficient.

This type of crane has occasionally proved quite economic for the construction of precast concrete buildings (fig. 20.2), large foundations of uniform shape, dry docks, etc., usually where working space is at a premium.

The portal crane has also found an application in the slightly different form of a beam launching crane in bridge deck construction.

Figure 20.3 shows the gantry being used to span two bridge piers for placing the deck beams. Figure 20.4 illustrates the alternative use of the gantry as a bridging beam to provide temporary support for precast concrete bridge units raised from barges below.

Figure 20.2 Portal crane application in high use construction

Figure 20.3 Gantry crane as a beam launcher

Figure 20.4 Gantry crane as a temporary support in bridge construction

Chapter 21   HOIST

Before the almost universal introduction of cranes, the hoist was normally the device selected to carry out the task of raising materials above ground level. Today the method is mostly restricted to uses in building construction, such as low-rise housing and multistorey structures, for handling small and light loads of less than 1 tonne, e.g. bricks, mortar, roof tiles and similar building materials. The hoist has maintained a presence in civil engineering work, in the form of a passenger/materials elevator to transport personnel and materials quickly on massive structures, such as dams, power station buildings, chimneys, cooling towers, etc. It can also be used to supply concrete to supplement a tower crane in high-rise construction. In this latter situation, the rate of concreting may be increased by using the hoist skip to deposit concrete in a wet hopper at the working level. The hopper in turn feeds the tower crane skip, thus allowing more time for the actual concrete placing, by reducing the lifting period. In this manner at heights of about 60 m, production can be stepped up by over 50%. Even so, both in high- and low-rise buildings, a combination of the tower crane and concrete pump is often preferred with the trend for low-rise construction more towards the use of the fork-lift truck, which is undergoing rapid development as a construction industry machine. However, for the construction of one to perhaps a dozen houses, the hoist is likely to remain economic.

The range of hoists available to suit different tasks is roughly divided into mobile and fixed hoists.

MOBILE HOISTS

These are light, mobile and robust and can be transported from site to site on a truck or towed. Once on site they are easily erected and may be operated by unskilled labour. The smaller models are used, for example, on housing where the operating height is less than 7 m.

The unit consists of a base frame, a winch unit powered by a small self-contained diesel engine, and two mast sections, which are folded over for transporting. The base incorporates four screw jacks to level up the unit when in the operational mode.

Figure 21.1
Arrangement for
transporting 1 tonne
portable materials hoist

Simple arrangement keeps maintenance costs low which combined with the low purchase price has ensured popularity with building firms.

A range of free-standing hoists is available for lifting materials weighing 250–1000 kg to landing heights of about 15 m.

The equipment has proved economical in the past, but suffers the disadvantage that materials must be transported to and from the landings by other means. Quite often this involves manhandling, and other forms of moving materials are now seriously eroding the hoist's advantages, most notably the fork-lift truck and concrete pumps.

FIXED HOIST

Fixed hoists are designed for operation in one location for a long period. They are most commonly used in high-rise construction, such as office blocks, chimneys, cooling towers, etc. In the case of multistorey buildings, it is often more economic to use a materials/passenger hoist after completion of the structural frame by tower crane, particularly where external scaffold is not involved, as the crane cannot place the materials at the required floor levels very easily. Also the construction of chimneys, lift shafts, cooling towers, sometimes involves concrete pumps and so tower cranes would not be required, the inclusion of a fixed hoist to transport labour, materials and equipment quickly, then becomes essential.

MATERIALS/PASSENGER HOISTS

These hoists are designed for loads of 500–1500 kg or about twenty passengers. In general the cage size is about 2.5 m high × 3 m wide × 1.25 m broad, which may be either a single front-mounted cage or two side-mounted cages (fig. 21.2). The most popular system for the majority of construction projects is a unit having a maximum operating height of 100 m with a cage of capacity 1000 kg or twelve passengers, and hoisting speed of 40 m/min. However, with the mast suitably braced at roughly 18 m centres such hoists can be operated at heights up to 600 m.

The cathead (i.e. sheaves at the top of the mast over which the winch rope passes from cage to winch), is mounted independently on the vertical mast sections allowing the tower to be extended whilst the hoist remains operational. As no counterweight is employed the cathead may be raised without disturbing the hoist reeving. The extension sections are raised into place by a small jib powered from an electric motor housed on the cage as shown in fig. 21.3.

Offering similar advantages a rack-and-pinion mechanism provides a popular alternative system particularly for access platform applications in fixing cladding, etc, on high-rise construction (fig. 21.4).

Figure 21.2 Fixed-position hoist – single and double cages

HOIST DATA

Capacity	500–1500 kg
Cage speed	40–90 m/min
Electric winch	25–50 hp (18–38 kW)
Voltage/frequency	380/50 V/Hz

Figure 21.3
Materials/passenger
hoist

erecting boom

cathead

concrete hoist

passenger and
materials hoist

mast

guide

safety entrance

to winch

hard standing

Figure 21.4 Access
platforms

CONCRETE HOIST

The concrete hoist is proving to be a very suitable supplement to the tower crane for concreting on the construction of high-rise structures. The hoist consists of a concrete skip, fixed either inside the mast or externally like the materials/passenger cage (fig. 21.3). When the required height is reached the concrete is discharged, sometimes automatically, into a storage hopper (wet hopper), and then distributed by wheelbarrows or transferred into the crane's concrete skip. Skips are available in the range $\frac{1}{2}$–1 yd³ (0.38–0.76 m³) capacity, the larger version is designed to carry 2000 kg at speeds up to 100 m/min, or heavier loads more slowly, with an electrically powered winch of about 100 hp (76 kW). Ideal output data are given in fig. 21.5 for a 1 yd³ (0.76 m³) capacity skip. The production is likely to be reduced in practice due to general factors limiting the efficiency.

Figure 21.5 Production data for a concrete hoist

HOIST EXERCISE

The construction site shown in fig. 21.6 illustrates the layout of building materials and access roads for the purposes of servicing two hoists for the erection of a low-rise building. Comment on and criticise the present layout in relation to the positioning of both the materials and hoists.

Suggested Solution

(A) Criticism of existing site layout (fig. 21.6):
    (a) Both hoists have separate scaffold staging, causing increased costs.
    (b) Materials are not stockpiled near hoists.

    (c) Entrance to the site is too narrow for trucks to pass.

    (d) Stores are located behind the batching plant so obscuring storeman's view and the check point is separated from the stores.

    (e) Concrete and mortar mixers are located too far from the hoists, and some distance from the temporary road.

    (f) Stockpiles are dispersed and hinder unloading and policing.

    (g) Temporary roads are long and narrow.

    (h) Some stores areas are difficult to reach, and cannot be seen from stores.

(B) Suggested improved layout (fig. 21.6):

    (a) Both hoists are housed in a common scaffold.

    (b) Batching plants have direct discharge into dumpers onto the hard-standing.

    (c) The temporary road is shorter and wider.

    (d) The access has been widened near the site entrance.

    (e) The stores is located to give a good view of all materials stockpiles, and is sited near the temporary road. The check point is also near the stores.

    (f) Aggregate bins can be directly accessed from the hard standing.

    (g) A compound is provided to police non-bulk materials.

Figure 21.6 Example of hoist site layout

Chapter 22   FORK-LIFT TRUCK

Since the early 1960s the fork-lift truck has made a great impact on materials handling in the construction industry and in particular on general building. On many sites, such as housing estates, much time and effort is wasted in moving materials from the initial delivery point to the subsequent places where they are required. Deployment of the fork-lift truck provides both a transport and lifting facility and so obviates the need for a hoist or even cranage of any kind. Even when a hoist (and therefore dumper trucks also) is required, the fork-lift is often a better choice than the dumper truck, especially as many materials are increasingly being packaged to suit fork-lift handling. Today the fork-lift truck has the ability to transport horizontally, and vertically at up to 10 m and to raise loads of several tonnes, although for construction site work trucks within the range 1000–5000 kg are more typical. Some machines now have forward reach capabilities and other sophistications as shown in fig. 22.1.

FORK-LIFT TRUCK ASSEMBLY AND OPERATION

The machine comprises a chassis supporting a diesel engine over the rear axle and a lifting mast at the front. The machine can be supplied with two- or four-wheel drive. In the two-wheel drive arrangement the drive is usually applied through the front wheels via a torque converter and power shift gearing. Both the steering and mast operation are hydraulically controlled. For construction work the machine is robustly constructed and supplied with large wheels and correspondingly large, wide tyres to cope with the rough and uneven terrain.

The advantage of the vehicle is that it can unload from any side of a delivery wagon. The forward extending truck in particular can often reach across the full width of the vehicle, for example where limits are imposed on working space on one side.

Once the materials have been unloaded they may be stacked in convenient stockpiles or transferred directly to the place of work, e.g. bricks to scaffold levels, thus avoiding double handling.

The load is carried as near to the ground as possible to lower the centre of gravity and reduce the risk of overturning, the load only being raised when

Figure 22.1 Fork-lift
truck options

(a) STANDARD   (b) FORWARD REACH
    ARRANGEMENT   CAPACITY

(c) LIFTING BEAM   (d) SHOVEL LOADER   (e) CONCRETE SKIP
                       (approx. 1m³)   (approx 1m³)

the discharge point is reached. Because of the fairly uneven terrain on the construction site the mast is equipped with hydraulic rams to move the load radially and thus avoid the need for too much manoeuvring of the vehicle itself.

Increasingly the telescopic boom version of the machine (fig. 22.2) is proving popular and facilitates loads to be placed well inside a building at heights up to 10 m.

Figure 22.2 Telescopic
boom fork-lift truck

Rapid developments are also taking place with extended booms for access platforms. Typical applications include cladding, painting, street lighting maintenance etc. The demarcation of forklifts and mobile cranes by different manufacturers for these aspects is becoming increasingly blurred.

STANDARDISED LOADS

The fork-lift truck requires standardised loads to operate economically. This is being increasingly assisted by the specification of standard components for building work, e.g. bricks, blocks, roof tiles, paving slabs, kerbstones, timber, joinery, bagged cement, roof trusses, drainage pipes, lintels, scaffolding tubes, packaged in economic quantities to reduce breakage and wastage. Many of these items, furthermore are delivered on wooden pallets at the factory, which are later returned when freed at the construction site.

Chapter 23

GENERAL POINTS ON THE SELECTION AND USE OF LIFTING EQUIPMENT

COMPARISON OF ALTERNATIVE METHODS OF MATERIALS HANDLING

HOIST

The hoist is a cheap and rapid method of lifting materials, but it does not load, stack, carry or place materials and usually the materials need additional transport by some other means, such as the dumper plus manhandling.

TOWER CRANE

The tower crane will off-load from delivery vehicles, lift, carry and stack, place and pour concrete, etc., but requires a banksman/dogman and is slow to manoeuvre. It tends to lose its advantages over the fork-lift truck for low-rise building construction such as housing except for placing materials inside the structure. It is, however, ideally suited to foundation construction and engineering work, and high-rise construction because of its clear span of outreach.

MOBILE AND TRUCK CRANE

These have similar characteristics to the tower crane, but greater lifting capacity. Firm ground, good access roads and clear unobstructed outreach are required.

FORK-LIFT TRUCK

The fork-lift truck will off-load, lift, carry and stack and does not require well-prepared roads, but can only lift around the sides of the structure. It is limited to heights of about 10 m.

BOOM SELECTION FOR CRANES

CANTILEVER-BOOMS

Cantilever-booms give greater clearance where headroom is limited, particularly when handling bulky loads. Lifting capacity is generally lower than for a comparable size strut-boom crane, because of the need to carry self-weight and for permanent site based cranage, the crawler strut-boom crane tends to be more economical for construction work duties, the cantilever being used mainly on wheeled vehicles in stockyards where manoeuvrability and headroom are at a premium.

STRUT-BOOMS

Strut-booms are used with derricks, crawler and truck-mounted cranes. They have high lifting capacity with long reach but are cumbersome and require dismantling to increase the boom length or to add fly-jibs. Dragline and grabbing crane, duties are suitable options.

FLY-JIB

A fly-jib may be attached to either a strut- or cantilever-boom to provide extra reach in the offset position. The arrangement is particularly useful for placing concrete reinforcement, and other light loads in restricted situations.

TELESCOPIC-BOOM

The telescopic-boom is a variation of the cantilever, offering a quickly operational and variable length boom, for mobile and truck-mounted cranes but cannot normally handle dragline and grabbing duties. The self-weight of the telescoping rams reduces the weight/radius capacity compared to the strut-boom, but offers increased manoeuvrability. It is an expensive type of crane.

BOOMS AND SAFE LOAD INDICATORS

To prevent the driver of a crane exceeding the permitted working load under normal working conditions, a safe load indicator is installed and calibrated to warn of overload on the hook.

The signalling device takes the form of a coloured flashing light and a bell mounted inside the driver's cab.

The indicator mechanism itself is usually mounted on the boom and must be fitted to all cranes of 1 tonne lifting capacity or over. There are several types of system and equipment approved according to the particular country and safety standards in operation.

LIFTING ATTACHMENTS FOR CRANES

Figure 23.1 shows the commonly used devices attached to the lifting hook of cranes of all types.

Slings (a) are used widely for moving bundles of reinforcement and timber, steel and concrete beams, precast concrete units, etc. The safe working load for two-legged slings is generally quoted at 90° as the lifting capacity varies with the angle of the legs. For example, a two leg wire sling of 19 mm diameter will carry 5280 kg at 60°, 4300 kg at 90° and 3050 kg at 120°. The capacity of the equivalent single leg sling is 3050 kg.

Lifting beam (b) – sometimes it is preferable to design and fabricate a steel beam for lifting slender and fragile items, e.g. precast concrete beams.

Lifting brackets (c) and *pincers* (d) – increasingly materials are delivered as standard packaged items, e.g. bricks banded in units of 1000. Specially designed brackets can usually recoup the initial fabrication costs from the reduced unloading time.

(a) Slings

(i)

(ii)

(c) Lifting bracket for packaged materials

Figure 23.1 Crane lifting attachments

(d) Pincers

(b) Lifting

CONCRETE SKIPS

Figure 23.2 shows concrete skips. The bottom dumping skip (b) is selected for mass concrete work where positioning is not critical. For placing concrete in thin walls, beams, columns, etc., the side-dumping skip (a) is preferred.

Figure 23.2 Concrete skips for application with a crane

(a) SIDE-OPENING CONCRETE SKIP

(b) BOTTOM-OPENING CONCRETE SKIP

(c) TIPPING CONCRETE SKIP

Sometimes the tipping skip (c) is preferred. It is filled from dumper trucks standing at ground level, thus the skip can be loaded whilst a second full skip is hooked on to the crane.

SAFETY REQUIREMENTS

It is essential to appoint a banksman/dogman to guide the crane operator and unhook loads, etc., especially when operating with restricted vision, such as tower cranes engaged on high-rise construction. The accepted signals are shown in fig. 23.3.

Figure 23.3 Banksman signals for crane operations

1 ALL MOVEMENTS
2 LUFFING OR DERRICKING BOOMS
3 TELESCOPING OR TROLLEY BOOMS
4 ALL CRANES
5 MOVING CRANES
6 ALL CRANES

Chapter 24   AERIAL CABLEWAY

The cableway is a form of overhead crane, with a cable substituted for rigid girders, so permitting a greatly increased span. For this reason, it has found successful application for the construction of concrete dams, in particular. The cableway has also proved ideal for the construction of river barrages, bridges and viaducts, and is occasionally economic for handling materials in quarries and stockyards, but the gantry crane tends to be favoured for these latter duties.

CABLEWAY HOISTING AND ROPING SYSTEMS

The simplest roping arrangement shown in fig. 24.1 consists of a track cable firmly anchored at each end to heavy steel towers. This cable provides the support for a wheeled carriage, made of a light frame containing two wheels for running on the track cable and two further sheaves to guide the hoist line. The carriage is winched between the towers by a looped travel rope which is played out and wound in simultaneously, causing the carriage to ride along the track line. The hoist rope on the carriage is independently operated from a hoist winch, the line is fed from the winch over a sheave in the head tower, to the first sheave on the carriage, down and under the hook block sheave, up and over the second carriage sheave and is finally anchored to the tail tower. The load may thus be raised and lowered at any position between the towers.

Figure 24.1 Simple cableway roping method

179

Figure 24.2 The
Lidgerwood roping
method

Figure 24.2 The Lidgerwood roping method

This method is generally only practical for short spans as the hoist rope tends to sag and get entangled with the other lines but the arrangement allows high travel and hoisting speeds and is consequently quite economical.

To overcome the roping problems on long spans Lidgerwood in the USA developed a system of rope carriers (fig. 24.2). In this method, an extra cable is provided fitted with buttons spread out along its span between the towers. The buttons are of increasing size working away from the head tower. As the carriage moves from the head tower a rider is trapped at each of the buttons. The amount of sag in the hoist and travel ropes is thereby significantly reduced, as the self-weight of each rope is transferred on to the track cable through the riders, which also relieves the hoist and travel winches from the need to cope with the dead weight of sagging cables. In this way the possibility of the cables becoming entangled is much reduced.

Figure 24.3 Button rope and rider used by the Lidgerwood roping method

This Lidgerwood design, however, requires the hoist line to be 'dead-ended' on the carriage rather than at the tail tower, since the riders are gathered in on the return trip to the head tower. The hoist drum is therefore arranged to pay out and wind in, in co-ordination with the travel of the carriage.

TYPES OF CABLEWAY

The aerial cableway is an expensive item of equipment whose capital cost can only be justified if the depreciation extends over the full duration of the contract. As this is unlikely, construction firms usually specialise in the type of construction work where specific types of cableway are required so that continuity of work is available. Cableways are made in a variety of types, custom engineered to suit the required conditions of the construction site. The more common types are the fixed cableway, luffing cableway, radial travelling cableway and parallel cableway.

FIXED CABLEWAY

The head and tail towers are held in a fixed position as shown in fig. 24.4 and may be sturdy framed structures or simply light guyed masts depending upon the span and loads to be carried. Only a straight line area may be serviced and fixed cableways are consequently used mainly for transporting materials over rivers and ravines rather than for construction work.

Figure 24.4 Fixed cableway

LUFFING CABLEWAY

For the construction of relatively long and narrow structures, such as bridges, the luffing cableway is sometimes suitable. The head tower is fixed, while the tail mast is mounted on a ball pivot base so that the mast may sway up to 20° from the vertical. The mast is guyed loosely at the side and tied firmly at the rear, thereby allowing the cableway to cover the radial area circumscribed by the luffing movement.

Typically a system with 30 m high masts should, at the widest point, serve an area 9 m times the length of the span.

Figure 24.5 Luffing cableway

RADIAL CABLEWAY

A typical arrangement is illustrated in fig. 24.6. The head tower is secured and counterweighted at a fixed position, whilst the tail tower or mast is mounted on a travelling carriage on rail tracks, which are set radially about

Figure 24.6 Radial moving cableway

the head mast. The angle of travelling is usually less than 40°. For angles greater than 20°, the head mast must have two stays to cope with the high horizontal force in the cable acting unsymmetrically through the tower. A 40° traverse provides a significant improvement upon the luffing-boom, and makes the system most suitable for the construction of long and relatively wide dams and bridges.

PARALLEL MOVING CABLEWAY

Where a cableway system is chosen to construct a very wide dam, then the only alternative is the fully moving cableway, whereby the head and tail towers are both mounted on rails and travel parallel to each other. The system is cumbersome and requires two operators to co-ordinate both units. A more economic arrangement can sometimes be achieved by setting up a pendulating mast (fig. 24.8).

Figure 24.7 Parallel moving cableway

Figure 24.8
Pendulating mast

CABLEWAY FEATURES

WINCHES

The winch is mounted on the head tower, but the controls are positioned so that the operator can obtain a clear view of the work. Sometimes on dam construction, as the dam height increases the driver's view gets obscured and radio communication is then necessary.

The winches are electrically powered with speed-regulated motors.

OUTPUT DATA

Tower travelling speeds (on rails)
Carriage speed	V_T	up to 12 m/min
Raising load	V_c	up to 360 m/min
Lowering load	V_R	up to 100 m/min
	V_L	up to 170 m/min

The maximum span of a cableway is dependent upon the tensile strength of the wires of the cable, the weight of the cable and the permissible amount of sag. With 75 mm diameter cable of breaking strength of 500 tonnes, a 5 tonnes load could be handled on a span of 1500 m. Generally, however, spans are within the range of 60–1000 m.

WORKING LOADS

For construction work and particularly when placing concrete in dam building, typical loads are 3–10 tonnes. Cableways have been designed for 10–20 tonnes and even 50 tonnes on a 1000 m span, but usually 25 tonnes is the practical limit with spans up to 900 m.

CABLEWAY CYCLE TIME

Figure 24.9 Cycle time for cableway

The typical activities involved in cableway operation are shown in fig. 24.9 as follows:

(1) Fill and attach concrete skip $t_1 = 40-60$ s

(2) Raise load $t_2 = \dfrac{L_T}{V_R} + 10$ s

(3) Carriage travel $t_3 = \dfrac{L_C}{V_C} + 10$ s

(4) Lower load $t_4 = \dfrac{L_W}{V_L} + 10$ s

(5) Empty skip $t_5 = 20-30$ s

(6) Raise skip $t_6 = \dfrac{L_W}{V_R} + 10$ s

(7) Carriage travel $t_7 = t_3$

(8) Lower skip $t_8 = \dfrac{L_T}{V_L} + 10$ s

When the carriage travel and load hoisting can be performed at the same time, then the cycle time shown in fig. 24.9 is reduced to

$$\text{cycle time} = t_1 + t_5 + \text{max. of} \quad \begin{matrix} t_2 + t_4 + t_6 + t_8 \\ \text{or} \\ t_3 + t_7 \end{matrix}$$

L_T = hook height above loading point;
L_W = hook height above discharge point;
L_C = distance between load point and discharge point.

In general, for concreting work the daily output is likely to reach only 60% of the theoretical and over a month perhaps only 50%. For other types of work, such as handling precast concrete units for bridge decks output is likely to be much lower.

Chapter 25 BELT CONVEYOR

The belt conveyor is proving economical for the transport of large quantities of material over long distances. In the past the belt conveyor was mainly associated with tunnelling and ore extraction in the mining industry, but in recent years this method of materials transport has been offered to the construction industry as an alternative to trucks and rails, for example in the placing of concrete for the construction of the foundations of massive industrial plants.

BELT CONVEYOR ASSEMBLY

Figure 25.1 Belt conveyor arrangement

The belt conveyor consists of an endless flexible, moving belt supported intermittently by idler rollers (fig. 25.1). The belt is driven from a head drum by an electric motor, and returns around a tail drum. Slack in the belt is taken up by a tensioning device, either from the tail drum or with a vertical gravity tensioner as shown in fig. 25.2.

The whole arrangement is supported on a light steel frame.

THE BELT

The belt must be strong and hard wearing. The most common form consists of plies of cotton/nylon duck (sacking) vulcanised with rubber to provide strength in tension. The top and bottom surfaces of the belt are covered with

185

Figure 25.2 Belt
tensioning device

Troughed load carrying
side of belt

Return
side of belt

Drums

Supports

Return drum with
weighted box attached

a layer of rubber, PVC or similar material to protect the belt from abrasion
and impact loading.

The thickness of the outer layer depends upon the type of material to be
carried, as shown in table 25.1 for a rubberised surface.

Table 25.1

Material	Top surface thickness (mm)	Underside surface thickness (mm)
Sand	2	1
Aggregate	4–5	2
Ores, sharp stones	5–8	2–3
Maximum	16	8

The belt is formed into a trough shape to increase the carrying capacity,
and reduce spillage (fig. 25.3). Before the introduction of cotton/nylon and
cotton/terylene fabric belts the idlers were set at 20° to 30°, but these new
belt fabrics now provide sufficient flexibility and robustness to allow angles
up to 55°, thereby significantly reducing spillage.

5 roller sets used
on belts over
0·7m wide

load carrying
side of belt

30° troughing
idlers

windboards
each side of
belt

return side
of belt

flat return
idler

Figure 25.3
Arrangements of belt
troughing idlers

The spacing of the idlers along the belt depends upon the weight of material to be carried together with the belt construction and troughing angle. The return idlers can be spaced more widely and act mainly to reduce belt sag.

GENERAL DATA

(a) *Idling rollers* – generally the carrying idlers are set at 1.0–1.5 m centres which close to 300–400 mm near the loading point. For the return idlers, the centres are approximately 2–3.5 m. The roller diameter is typically 50–250 mm, depending upon the belt width.

Figure 25.4 Two-motor driven conveyor belt

(b) The size of the drive motor depends upon the length and width of the belt and the density of material carried. For very long lengths and heavy material two motors or more may be required, as shown in the arrangement in fig. 25.4.

(c) *Gradient* – most modern belts are capable of holding the material in a stable condition at an inclination of between 15° and 25°.

(d) *Drive and tail drums* – the distance between these two drums defines the operating length on the conveyor. For most uses the design is tailor made to the specific application and the length of conveyor and can range from a few metres up to 10 km or more. One of the longest systems in use is 100 km, assembled in 11 km sections.

(e) Belt widths vary from 400 to 3000 mm, with 1 m as the typical requirement.

OUTPUT

The output from a conveyor belt is given by the equations:

$$Q_v = 3600 \times V \times A$$

or

$$Q_t = 3600 \times V \times A \times D$$

where Q_v = theoretical production in m³/h;
   Q_t = theoretical production in tonnes/h;
   V = belt speed in m/s;
   A = cross sectional area through material on the belt in m²;
   D = material density in tonnes/m³ (i.e. loose density).

The loose volume carried per metre on the belt is governed by the width of the belt, which is itself influenced by the particle size of the material being carried.

The theoretical outputs for specific band widths are shown plotted against band speed in fig. 25.5. A bandspeed of about 1.5 m/s is typical.

For coarse material of particle size 50 mm or more the band width required is approximately as follows:

band width = 4 × particle size + 15 cm (uniform lumps)
band width = 2 × particle size + 20 cm (non-uniform lumps)

For non-uniform lumps particle size is taken to be the greatest lump dimension.

Figure 25.5 Production data for a conveyor belt

output m³/hr

band width (mm)

Band speed (m/s)

PORTABLE CONVEYOR

Figure 25.6 Portable conveyor

For moving materials on the construction site, flexibility is required. The expensive and almost permanent conveyor system previously described is unsuitable except for tunnelling or similar work. The portable conveyor however, is a relatively inexpensive, and versatile device which can be of use on many construction sites, where the transport or disposal of bulky materials is required.

The arrangement shown in fig. 25.6 consists of a portable frame supporting a conveyor belt, rollers and electric motor. The length of the conveyor is usually of fixed length, about 3–6 m, but models are available up to 20 m.

Applications are constantly widening, although as yet not a common item on the construction site, where the flow path requires regular repositioning.

(a) Materials movement is independent of surface shape, form and condition.
(b) The conveyor can be used on poor load-bearing soils, over road junctions, streams and even small rivers, where other methods would be inconvenient and more expensive.
(c) A conveyor requires little labour and maintenace.
(d) Transporting materials by conveyor is relatively independent of the weather.

However, large and sharp, fragmented material such as broken rock tends to damage the belt. Also where a long conveyor is installed a breakdown brings the whole system to a halt. Although this can be avoided where portable units are used.

HANDLING OF CONCRETE

Concrete conveyors are provided with narrow belts of 300–400 mm width thereby reducing the load per unit length is reduced compared with the standard conveyor. The unit is as a consequence light, easy to install and manoeuvre. High belt speeds 2–3 m/s are typical, which assists in keeping the belt clean as the concrete is thrown off on reaching the head drum, often augmented with a scraper (fig. 25.7) or brush (fig. 25.8).

Figure 25.7 Belt cleaning method with a brush

Figure 25.8 Belt cleaning method with a scraping tool

CONCRETE DELIVEY ON A 400 mm Wide Belt

Belt speed (m/s)	Delivery rate (m^3/h)
2	50
3	115

These delivery rates compare favourably with concrete pumps and only suffer the disadvantage that the gradient is limited to 15° or so. The problems of quick positioning are also being overcome, for example by means of the truck-mounted telescopic boom, whereby two or more conveyors are telescoped out.

CONVEYORS FOR EARTHWORKS

Usually the most economic method of moving excavated earth on the construction site is by scraper, loader shovels, trucks or lorries. However,

Figure 25.9 Conveyors for application on excavation tasks

189

Figure 25.10
Conveyors for
application on
earth placing tasks

research work has been carried out to asses the feasibility of using a conveyor system in conjunction with large draglines and bucket excavators on, for example, motorway construction (fig. 25.9) and alternatively bringing fill over a long distance to be spread by conventional earthmoving plant (fig. 25.10).

Chapter 26   MONO-RAIL

Figure 26.1 Mono-rail materials handling system

The mono-rail, before the advent of reliable concrete pumping methods, was used extensively on sites where concrete or other materials needed transport over short distances through congested areas where trucks, etc., could not pass. Today the method is perhaps more of an alternative to a conveyor system and is probably still preferred because of the flexibility allowed in stopping the wagons wherever required.

The transporting skip consists of a self-travelling tippable car, which drives on a single track by making contact with the top of the rail. The unit is steadied by means of four idler rollers spring loaded against the flange of the rail.

The car can be started by unskilled labour and halted at a selected position by a tripping mechanism.

The track is assembled from short sections of rail mounted on tressles, which either rest on a firm base or are bolted to structural steelwork, for example in concrete placing of a bridge deck, the system is used up to about 300 m, but this can be increased where the track is looped. The cars thereby are started at different times to maintain a steady concrete delivery rate.

The car is about $\frac{1}{2}$ m^3 capacity and will travel up an incline of 1 in 20 at 2 m/s.

EXERCISE

The exercise involves the generation of alternative methods of producing, transporting and placing concrete from the selection of equipment described in previous chapters. The aim is to encourage rational plant selection by assembling a method plan and costing out the various alternatives.

Problem

Your company has recently obtained a sub-contract for the construction of the concrete floor for a process factory as shown in fig. 26.2. As the planning engineer you are required to provide a method plan and statement together with a cost estimate for the production, transporting and placing of the concrete.

Figure 26.2 Materials handling plant section exercise site layout

Chapter 27 CONCRETE PUMP

Concrete pumps were first introduced into the construction industry in the 1930s, but until the development of hydraulic pumps in the 1960s the mechanical and maintenance problems were responsible for the general lack of interest. Today the concrete pump is accepted as a much easier, quicker and cheaper method of placing concrete compared with many of the other methods of handling.

MECHANICAL CONCRETE PUMPS

Figure 27.1
Mechanical concrete pump

The early pumps were fixed position, which had flexibility only at the discharge end of the pipe. Typically these were 150–200 mm diameter and consequently were too heavy and cumbersome for construction site use.

The pump (fig. 27.1) consists of a feeding hopper, inlet and outlet valves and piston, all driven by a fly-wheel. The method causes high wear and tear on the valves and a jerky flow of concrete. In addition the concrete mix must be rich and fatty, if blockages are to be avoided.

When the pump is operating satisfactorily the following outputs are possible.

Output	m³/h	12	20	40
Number of pistons		1	1	2
Pipe diameter	mm	125–150	150–180	150–180
Cylinder diameter	mm	160	195	195
Power	kW	20	35	45
Horizontal distance	m	300	300	300
Vertical height	m	40	40	40

HYDRAULIC CONCRETE PUMPS

The introduction of the hydraulic concrete pump has had the greatest bearing on the acceptance of concrete pumping methods.

The pumps may be either truck-mounted or towed as a separate unit. Oil is the working fluid, used to operate either single (fig. 27.2) or more commonly dual pumping cylinders.

Figure 27.2 Single-cylinder hydraulic concrete pump

Concrete is drawn from the hopper into the cylinder on the suction stroke, and discharged along the pipe on the pressure stroke. A diesel engine powers the hydraulic pump but occasionally an electric motor may be used.

This method of pumping provides several advantages over the earlier mechanical pumps as follows:

(a) The pump is designed with a long piston stroke, requiring fewer strokes with a consequent reduction in static friction encountered by the concrete.
(b) The pressure exerted on the concrete may be varied to control vertical pumping (fig. 27.3).
(c) The piston is accelerated at the start of the delivery stroke and then maintained at a steady speed, so producing a smooth and even discharge.

Figure 27.3

Relationship between oil pressure and concrete pumping height

The new dual cylinder models have the ability to deliver concrete at up to 120 m^3/h with a pipeline diameter of 150 mm. Such volumes, however, would cause handling difficulties on most sizes of concrete pour and the usual pipe bore is 100 mm diameter, giving a corresponding output of up to 35–45 m^3/h. Some models include speed regulation by varying the oil flow and pressure, allowing output to be adjusted down to 20–25 m^3/h or less. Hydraulic pumps have been used to deliver concrete, through over 600 m of pipeline to heights exceeding 300 m or so. Bends in the pipeline reduce the effective pumping distance by about 10 m for each 90° bend, 6 m for 45° and 3 m for 22½°.

PUMP OPERATION

(a) Before pumping is started a grout mixture should be flushed through the pump and pipeline, to provide lubrication. The best results are achieved by initially inserting a plug into the pipeline, which is then forced along with the grout.

(b) The concrete in the hopper should be agitated, to keep the concrete moving and the concrete level must be constantly topped up to prevent air pockets forming in the pipeline.

(c) A compressed air supply nearby is useful for clearing obstructions in the pipeline.

(d) Pumping tends to be most efficient with concrete having a slump of 50–75 mm, with a high percentage of fines. The design of pumpable

Figure 27.4 Rubber plug for cleaning pump delivery pipe

mixes is fairly specialised, however, and the reader is advised to seek further information, e.g. from the Cement and Concrete Association.

(e) To clean out the pipeline a section of pipe is removed and the hopper is pumped clean. A plug is inserted into the remaining pipe and forced along the line by compressed air or water pressure (fig. 27.4).

PUMP MOUNTING

TRUCK-MOUNTED PUMP

The demand for the occasional concrete pour such as in high-rise construction, bridge decks, etc., coupled with the advantages of ready-mixed concrete, has led to the development of truck-mounted concrete pumps for short-term hire, e.g. the duration of the concrete pour. Many sizes of mobile, truck-mounted hydraulic concrete pumps are now available, supporting booms with a reach of up to 50 m (fig. 27.5), depending upon the restrictions placed on axle loads for travelling on public highways.

Figure 27.5 Truck-mounted concrete pump

TRAILER PUMP

Figure 27.6 Trailer-mounted concrete pump

As the size of buildings has increased, many structures require access beyond the reach of truck-mounted boom pumps. The trend in such situations is to use a trailer concrete pump in conjunction with a portable hydraulic boom, fixed either to a mast (fig. 27.7a) or to the structure (fig. 27.7b).

Figure 27.7(a) Portable mast and boom

Figure 27.7(b) Portable hydraulic boom

SQUEEZE PUMP

A flexible rubber pipe secured half circle in an air-tight drum is squeezed by two diametrically opposed rollers successively flattening and squeezing the

Figure 27.8 Squeeze pump

concrete towards the outlet. The operation takes place in a vacuum at 0.8–0.9 bar and thus the pipe regains its shape allowing more concrete to be sucked from the mixing chamber.

The unit may be lorry-mounted, with the hydraulic pump to drive the rotor coupled to the truck's diesel engine or alternatively the pump may be towed and powered with its own diesel engine.

The system requires little preparation time and provides a continuous pumping action. However, only very workable concrete can be pumped. The pipe is limited to 75–100 mm diameter.

DATA

Pipe diameter	75 mm
Max. output	28 m³/h
Pumping length, or	130 m
Pumping height	40 m
Power	80(60) hp(kW)
Self-weight	6000 kg
Travelling speed	90 km/h

PNEUMATIC PUMP

The apparatus consists of a compressor, compressed-air cylinder and pneumatic concrete pump. Concrete is charged into the pressure vessel of capacity 250–1000 litre. The vessel is then sealed and compressed at 2.5–3 bar from the storage cylinder. The concrete is then in effect blown along the delivery pipe into a wet hopper.

Figure 27.9 Pneumatic concrete pump

Delivery pipes of 150–180 mm are typical and bends must be securely anchored to cope with the fast moving concrete (0.5–1.0 m/s). Approximately twenty to forty charges per hour are possible at the maximum delivery distance (about 300 m along the horizontal).

The compressor must be capable of supplying approximately fifteen times the volume of concrete to be pumped, e.g. 1 m^3/h of concrete = 15 m^3/h of compressed air.

CEMENT PUMP

Many modern concrete batching plants transfer dry cement from a storage silo into the mixing hopper.

The pump to execute the transfer is shown in fig. 27.10. The cement is drawn from the hopper on to a screw feed which is rotated at approximately 1000 cycles per minute and fed into a chamber. The cement particles are blown along the delivery pipe by compressed air and may be transported horizontally up to 1500 m and vertically to heights of 30 m. The delivery pipe diameter range is 30–200 mm and correspondingly requires 0.5–3 kW for each tonne/h of delivered cement.

Figure 27.10 Powder cement pump

BIBLIOGRAPHY: SECTION 1

1. Nunnally, S. W. (1986) *Managing Construction Equipment*. Prentice-Hall.
2. Havers, J. A. and Stubbs, F. W. (1971) *Handbook of Heavy Construction Equipment*. McGraw-Hill.
3. Caterpillar Tractor Company (1987) *Caterpillar Performance Handbook*. Peoria, USA.
4. Mead, H. T. and Mitchell, G. L. (1972) *Plant Hire for Building and Construction*. Newnes–Butterworth.
5 International Harvester Company of Great Britain *Basic Estimating*. London
6. General Motors Corporation (1987) *Production and Cost Estimating of Material Movement with Earthmoving Equipment*. Ohio, USA: Terex division.
7. Komatsu Ltd (1987) *Komatsu Specification and Application Handbook*. Japan.
8. Peurifoy, R. L. and Ledbetter, W. B. (1985) *Construction Planning, Equipment and Methods*. McGraw-Hill.
9. HMSO (1962) *Military Engineering*, Vol. 16, *Engineer Construction Plant*.
10. Illingworth, J. R. (1972) *Movement and Distribution of Concrete*. McGraw-Hill.
11. Higgins, L. R. (1979) *Handbook of Construction Equipment Maintenance*. NY, USA: McGraw-Hill.
12. Russell, J. E. (1985) *Construction Equipment*, Reston, Prentice-Hall.
13. Oberlander, G. D. (ed.) (1986) *Earthmoving and Heavy Equipment*. Proceedings of Conference by the Committee of Construction Equipment and Techniques. Construction Division ASCE Tempe, USA.
14. Fletcher, B. G. and Lavan, S. A. (1987) *Civil Engineering Construction*. Heinemann.
15. Harris, F and McCaffer, R. (1982) *Construction Plant : Management and Investment Decisions*. Granada.
16. Gardiner, L. (1979) *Mechanical Plant in Construction*. Godwin.
17. Dickie, D. E. (1975) *Crane Handbook*. Butterworths.
18. Tucker, U. V. (1980) *Earthmoving Plant*. Woodhead–Faulkner.
19. Douglas J. (1978) *Construction Equipment policy*. McGraw-Hill.

20. Englestone, F. N. (1983) *Insurance for the Construction Industry*. Goodwin.
21. Antill, J. M. and Ryan, P. W. S. (1983) *Civil Engineering Construction*. Angus and Robertson.
22. Wood, S. (1977) *Heavy Construction*. Prentice-Hall.
23. Schnexnayler, C. J. (1982) *Construction Equipment and Techniques*. ASCE. March.
24. Ratay, R. T. (1984) *Temporary Structures in Construction*. McGraw-Hill.
25. Alekseeva, T. V. *et al* (1985) *Machines for Earthmoving Work*. Balkeema.
26. Kogan, J. (1985) *Lifting and Conveying Machinery*. Haifa.
27. Christian, J. (1981) *Management, Machines and Methods in Construction*. Wiley International.
28. Day, D. A. (1973) *Construction Equipment Guide*. Wiley.
29. Harris, F. C. (1983) *Ground Engineering, Equipment and Methods*. Granada.
30. The Construction (Lifting Operations) Regulations 1961. HMSO.
31. The Construction (General Provisions) Regulations 1961. HMSO.
32. The Construction (Working Places) Regulations 1966. HMSO.
33. The Construction (Health and Welfare) Regulations 1966. HMSO.
34. The Health and Safety at Work Act 1974. HMSO.
35. *Construction Regulations Handbook*. Royal Society for the Prevention of Accidents.
36. *Supervisors Safety Handbook*. The Federation of Civil Engineering Contractors. Safety in Construction Work series published by HMSO.

SECTION 2   CONCRETE PRODUCTION AND EQUIPMENT

This section concentrates mainly on the equipment needed for mixing and batching concrete, transportation and placing methods having largely been covered earlier under materials handling. The chapters include:

SAND AND AGGREGATE PRODUCTION

Many types of construction involve the production of concrete in regions distant from readily purchasable sources of materials. Construction engineers and managers are then faced with the task of setting up local quarrying and crushing facilities. Crushing equipment, screening and washing processes are described together with a worked example illustrating the calculations and decisions required in producing particular sized materials from crushed rock.

CONCRETE EQUIPMENT

An introduction to concrete materials and technology is provided focusing on handling and batching concrete, concrete mixers, central mixing plants and production output information.

PRESTRESSED CONCRETE

Both the longline and post-tensioned methods of manufacture and assembly are described together with guidelines provided on quality control and selection of materials.

Chapter 28   AGGREGATE PRODUCTION

Aggregate is used for hardcore, asphalts and macadams, concrete, mortar, general fill, etc. The sizes of the constituent parts of the aggregate depend upon the grading required, for example 150 mm down to 3 mm can be obtained by crushing rock.

The rock itself is usually obtained by blasting operations and transported to the crushing equipment. Modern crushers are capable of handling large blocks of over 1 m maximum dimension, with the choice of suitable machinery depending upon the costs of reducing the size of the fragments at the quarry and the method of transport to the plant. Following crushing, washing and sorting into various sizes is commonly carried out.

Sand, unlike crushed aggregate, can be produced artificially, but is normally obtained from natural deposits which may be subsequently mixed to give the required gradings.

Specifications for aggregate types and gradings are beyond the scope of this work, as these depend upon the use. For help in selecting materials, codes of practice are available for concrete, road works, fill material, etc.

EQUIPMENT FOR AGGREGATE PRODUCTION

The capacity for the crushing plant is selected to produce a given rate of tonnage output. The equipment required is as follows:

1. Feeders and hoppers.
2. Primary crusher.
3. Secondary or reduction crusher.
4. Conveyors to transport the material between processes.
5. Screens and washing plant for separating, grading and directing the sizes to the bins.
6. Storage bins and discharge hoppers.

CRUSHING PLANT

The crusher takes the raw material and breaks it into small sizes. To do this the machine is set to the maximum size required. However, the rocks will fragment into a range of sizes down to dust, which can be fed away for further reduction or screening and sorting.

The machine is usually described in terms of its crushing ratio, i.e. the size of the feed opening to the setting for crushing (fig. 28.1). The range of crushing ratios for various types of machine are shown in table 28.1.

Figure 28.1 Jaw crusher

Table 28.1 Crushing ratios

Crusher	Smaller units	Larger units
Jaw	4–8	3–10
Gyratory	5–8	4–6
Impact	5–25	5–50
Cone	3–14	4–14
Twin-roller	1–4	2–8

It can be seen from fig. 28.2 that there are several different types of crushing machine, for handling different size feed and material. The output from each varies between types for a given crushing ratio and therefore

Figure 28.2 Crushing machine characteristics

205

careful costing is necessary when several stages of reduction are required. For primary crushing, i.e. breaking material down from large pieces, either a jaw or gyratory type is preferred while for secondary or further reduction, the cone, hammermill/or roll crusher is usually selected.

PRIMARY CRUSHERS

JAW CRUSHER

The jaw crusher is a simple but effective method, having a rectangular shape on plan and triangular from the side elevation. Three walls are fixed, while the fourth is a moving plate, so forming a jaw-type structure. Material is fed in at the top and passes out through the base. There are three types of jaw crusher, namely:

(i) Double toggle.
(ii) Single toggle.
(iii) Double jaw.

The crusher size is designated by the dimensions of the top opening.

Double Toggle Jaw Crusher

The machine was first developed by Blake in England and introduced in the 1850s. It is robust, durable, and works on the following simple principle. One side of the jaw (a) is held stationary, while the other (b) is attached to pivots at the top and bottom (c and d). The top pivot (c) simply acts as a suspension point, crushing motion being obtained by a rotating eccentric shaft raising and lowering the rod (e) connected by two toggles to the jaw (b) at (d) and a reaction point (h). The design principle is very effective with the crushing force increasing towards the top of the jaw. Also the opening and closing movement of the jaws allows small fragments to fall quickly through the bottom and not restrict the flow.

Figure 28.3 Double toggle jaw crusher

Single Toggle Jaw Crusher

This type of jaw crusher has only a single toggle at the bottom of the moving jaw connected to a reaction point. Motion is produced with an eccentrically rotating shaft, located at the top joint, so giving both vertical and horizontal movement, consequent greater attrition results in more fines than the Blake-type machine.

Double Jaw Crusher

A few machines have been made with both jaws able to move but have not proved popular.

Figure 28.4 Single toggle jaw crusher

Data for Jaw Crushers

Table 28.2 Dimensions, settings, capacities and power for portable jaw crushers

	Feed opening (mm)					
Item	1.1 × 1.2	0.9 × 1.15	0.75 × 1.05	0.65 × 1.00	0.5 × 0.9	0.45 × 0.8
Closed settings (mm)	125–300	100–250	120–200	75–200	50–180	30–180
Capacity (t/h)	275–750	180–600	125–575	100–400	40–250	40–130
Max. power (kW)	230	160	150	120	75	45
Weight (tonnes)	75	70	55	55	30	25
Max. output/max. power ratio	3.2	3.75	3.8	3.3	3.3	2.9

Grading Curve or Crushed Material

The crushed material will not be of uniform size, but graded. The proportions of each size depend upon the setting, type of raw material and crusher. Typical values for single and double toggle machines are shown in fig. 28.5 for hard non-brittle rock. The maximum size of material fed in should not exceed about 0.8 × the feed opening (fig. 28.1).

Figure 28.5 Grading curve obtained with a jaw crusher

Reduction Ratio

It can be seen from table 28.2 that for the jaw crusher the reduction ratio can be varied in the range 3–10 on a typical machine.

GYRATORY CRUSHER

The crusher comprises a hardened steel head mounted vertically on a shaft, all contained within a cast iron or steel housing, lined with hardened steel plates (fig. 28.6).

Figure 28.6 Gyratory crusher

The head has a long conical shape, with a through shaft suspended in a bearing at the top, and an eccentric base connection to gearing. Thus as the cone is rotated the gap between itself and the walling changes from a maximum to minimum each cycle. The rock is fed into the chamber at the top and as it moves downward, undergoes crushing, finally emerging through the bottom gap.

The crusher size is designated in terms of the top diameter of the housing D and the distance between the housing core at the same level, d. The setting, S, can be stated either for the maximum (open) or minimum (closed) gap at the base, but it is more usual to quote the larger value. The setting may be altered by raising or lowering the core or by introducing thicker wall plates.

Data for Gyratory Crushers

Table 28.3 Dimensions, settings capacities and power of gyratory crushers

Item	Feed opening (m) ($d \times D$)					
	1.4 × 4.0	1.0 × 3.6	0.75 × 2.5	0.5 × 2.0	0.4 × 1.5	0.33 × 1.1
Open settings (mm)	150–180	125–165	100–125	85–115	75–100	50–75
Capacity (t/h)	650–800	500–650	300–400	200–250	150–200	80–150
Max. power (kW)	225	210	130	95	75	50
Weight (tonnes)	45	45	40	35	35	30
Max. output/ max power ratio	3.55	3.1	3.1	2.6	2.6	3

Grading Curve of Crushed Material

Like the jaw crusher, the maximum size of material fed in should not exceed about 0.8 × the feed opening. The final grading curve is similar to that obtained from the jaw crusher. Figure 28.7 illustrates typical gradings for brittle and tough rocks, respectively.

Reduction Ratio

It can be seen from table 28.3 that for the gyratory crusher the reduction ratio (d/S) can be varied on a given machine, typically 4–6 for large crushers and 5–8 for smaller units.

Figure 28.7 Grading curve obtained with a gyratory crusher

CRUSHERS SUITABLE FOR PRIMARY AND REDUCTION CRUSHING

Figure 28.8 Hammermill

HAMMERMILL

The hammermill comprises hammers linked to a rotor or several rotors for a large unit, all contained in a strong housing (fig. 28.8). The hammers are rotated at high velocity, striking the lumps of material as they enter the chamber and breaking them up into smaller pieces. The base of the chamber is lined with a grating plate allowing grinding to take place and as a consequence more fines are produced compared with most other methods. The hammer size is designated similar to the jaw crusher in terms of the dimensions of the feed opening. The setting may be varied by adjusting the grate bars. Some variation in the production capacity can be obtained by altering the hammer velocity, e.g. 35–50 m/s circumferential velocity.

An alternative type of hammermill is the impact breaker (fig. 28.9). The material struck by the hammer is thrown against breaker plates, bars or grates situated above the rotor and so broken into smaller pieces. There is no bottom grating and therefore the amount of splintering is largely dependent upon the impact velocity, typically 15–30 m/s circumferential velocity. The size of the feed material should be much smaller than the feed opening and introduced into the chamber at a rate which will not cause clogging, and therefore initial screening is necessary to remove oversize pieces. The method is suited to soft rocks that are also not too abrasive, otherwise wear and tear will be excessive. Generally the impact breaker is used for primary crushing while the hammermill is preferred for reducing crushing duties. Typical data for a hammermill are shown in table 28.4.

Figure 28.9 Impact breaker

REDUCTION CRUSHERS

Material which has passed through primary crushing will generally have fragments distributed down from at least 50 mm in size even when small machines are used. Thus, for concrete, aggregate, etc., further reduction in size is required. Commonly material larger than a specified size is retained by screens after passing through the primary crusher, and subsequently reduced by further crushing in another machine. The crushers available for this purpose are cone crusher, roll crusher, rod mill and ball mill.

Table 28.4 Hammermill/impact breaker data

Item	Feed opening (mm)										
	225 × 325	275 × 535	275 × 830	300 × 850	300 × 1050	445 × 1100	445 × 1360	445 × 1625	445 × 1890		
Maxstone (mm)	250	75	100	100	150	150	250	250	250		
*Setting (mm)	2–65	2–65	2–65	2–65	2–65	2–65	2–65	2–65	2–65		
Capacity (t/h)	5–35	12–75	19–120	25–145	32–190	41–250	51–320	60–300	20–440		
Max. power (kW)	45	95	152	190	228	266	380	456	532		

* Hammermill only.

Cone/gyrosphere

The cone/gyrosphere crusher operates on a similar principle to the gyratory crusher described earlier, and is designed to produce large quantities of fine aggregate. The crushing head is much broader but shallower than the gyratory crusher and is connected at the base through eccentric bushing and gearing to the drive and rotated at a high rate of revolutions. The machine is designated in terms of the maximum diameter of the crushing head (cone), but it is also customary to quote the feed opening dimension. The cone crusher has straight sides to the head while the gyrosphere tends to be more curved. The setting may be altered by altering the relative position of the housing rings and head plates.

Figure 28.10 Cone crusher

Data for the cone crusher

Table 28.5 Dimensions, settings, capacities and power of cone crushers

Item	Max. cone dia. (mm)			
	600	900	1200	1650
Feed opening (openside) (mm)	90	130	190	280
Minimum discharge setting(s) (mm)	6–25	10–40	13–50	20–65
Capacity (t/h)	13–50	28–105	63–210	160–470
Max. power (kW)	20	50	75	185
Max. output/max. power ratio	2.5	2.1	2.8	2.5

Grading curve

Figure 28.11 illustrates a typical grading curve for the cone/gyrosphere crusher, and has a similar shape to that obtained for the gyratory machine used for primary crushing.

Reduction ratio

Table 28.5 indicates that for cone/gyrosphere crushers the reduction ratio can be varied on a given machine between 4–14 approximately.

Roll Mill

Roll mills are used for similar duties to the cone crusher and produce fine aggregate or even coarse sand (particle size less than 3 mm). The machine

Figure 28.11 Grading curve obtained with a cone crusher

Figure 28.12 Roll mill

comprises two counter rotating rollers, and material is fed through the gap between the two for crushing. Usually one roller has a fixed axis while the other can be adjusted to give the required setting, it is also spring loaded to prevent damage when oversize non-crushable material is encountered. Usually the roller surface is smooth, but corrugations on one or both are available depending upon the material for crushing. The maximum size of material that can be put through the system is proportional to the roller diameter. If pieces are too large they cannot be nipped by the rollers and would simply fall out. The maximum size stone, B, for input can be approximately determined from the formula: $B = 0.085 \times$ roller diameter + setting (all in inches).

The crusher is usually designated in terms of the setting, the diameter and length of the rollers.

Grading curve
The distribution of particle sizes is similar to that obtained from jaw or cone crushers. Course sand can be obtained from the lowest setting.

Reduction ratio
Table 28.6 indicates that the reduction ratio for a typical roll mill can be varied from about 1–4 for small crushers and 2–8 for larger versions. A comparison of roll and cone crushers (table 28.6) shows that for a given output/power ratio the cone crusher accepts larger size feed material and has a greater crushing ratio. Therefore fewer crushing stages would be necessary for the same size of input material.

Table 28.6 Dimensions, settings, capacities and power for roll mills

Item	Roller diameter × roller length (mm)						
	400 × 400	600 × 400	750 × 450	750 × 500	1000 × 500	1000 × 600	1400 × 600
Setting (mm)	5–60	5–60	5–60	5–60	5–60	5–60	5–60
Capacity (t/h)	15–140	15–140	15–160	20–200	20–200	20–200	20–240
Max power (kW)	25	25	55	75	75	75	115
Roller speed (rev/min)	120	80	60	60	50	50	40
Max. output/ max. power ratio	5.5	5.5	2.9	2.7	2.7	2.7	2.1
Max feed size (mm)	20–75	30–85	35–90	35–90	45–100	45–100	65–120

Ball Mill and Rod Mill

These units are mainly used to produce uniform sand from 15 mm size aggregate (approximately). The rod mill consists of a rotating drum containing 25–50 mm diameter rods occupying about 30% of the total volume. They are slightly shorter than the length of the cylinder. The aggregate is fed in at one end of the drum, broken down into small grains as it is rotated and crushed by the rods, finally emerging at the other end through a sieve. The feed material can be introduced wet or dry.

An alternative arrangement is to use steel balls instead of the rods (fig. 28.13).

Output: 1–200 m³/h
Power: 2–3 kW/m³ per h

Figure 28.13 Ball/rod Mill

SCREENING OF AGGREGATES

The process of sorting aggregate into different sizes is illustrated in the example shown in fig. 28.14. It can be seen that before each stage of crushing material smaller than the crusher setting is allowed to pass through one or more screens. In this way the capacity of the crushing units can be kept smaller than the actual quantity of material passing through the whole system. Commonly double or triple deck screens are used except at the primary stage where grizzle bars or scalping screens are needed to cope with large lumps of rock. Conveyors transport the crushed material between stages or to stockpiles.

Figure 28.14 System for crushing and screening

APRON FEEDERS AND GRIZZLES

Material for the primary stage of crushing is usually delivered in very large lumps depending upon the quarry and mode of transport to site. Thus robust equipment is needed to receive the dumped rocks and feed them at a steady rate into the crusher (fig. 28.15). Two types of feed are commonly used:

(i) Vibrating plate feeder.
(ii) Apron feeder.

Figure 28.15 Mobile feeder unit

Vibrating Plate

The feeder pan comprises a heavy steel plate, side liners and hopper all mounted on heavy duty coil springs. The vibrating unit consists of two parallel counterweighted offset shafts geared together (fig. 28.16b). Vibrating motion is produced by the interaction of the forces induced by the rotating shafts. In positions (2) and (4) all forces are acting against each other while for (1) and (3) the forces are in the same direction. Thus is these conditions the motion is up or down, and sideways, with one revolution producing a straight line reciprocating motion at 45° angle to the plane of the plate. Material is therefore alternatively lifted and conveyed forward.

This type of system is commonly selected for material obtained from a gravel pit.

Figure 28.16(a)
Vibrating plate feeder

FEED END

CRUSHER END

Figure 28.16(b)
Vibrating plate
mechanism

Apron Feed

The apron feed is necessary when quarried material is involved. The equipment is constructed to withstand heavily dumped loads and consists of a series of steel pans forming a continuous chain supported by steel idler rollers. The 'belt' is drawn by a head roller linked through worm reduction gearing to an electric or diesel motor. A variable rate of feed is available on most models through gear changes, but in any case at least 25% more capacity than that of the crushing unit is desirable.

Apron feeders are manufactured for a range of duties commonly described as standard duty, heavy duty and super heavy duty.

Figure 28.17 Apron feeder

Grizzles (scalping unit)

A grizzle is usually placed at the end of the apron feed and used to screen out material smaller than the crusher setting or if constructed as a double tier unit the top deck may be arranged to direct oversize material away from the

SINGLE STEP
VIBRATING GRIZZLY
FEEDER

Figure 28.18 Grizzle/
scalping units

crusher. The equipment is generally referred to as a 'scalping unit' and comprises a spring-mounted frame with heavy duty parallel bars forming the top deck and a tough perforated plate for the lower deck. The unit usually has a vibrating mechanism and is placed at a slight angle downwards towards the crusher. The material simply passes along the bars and is discharged at the end of the run into the crusher while the lower deck feeds directly on to other crushing stages (fig. 28.14).

Table 28.7 Vibrating plate feeders/grizzles

Item	Size (length × breadth) (m)				
	3.5 to 4.8 × 0.9	3.5 to 4.8 × 1.0	3.5 to 6.0 × 1.2	4.8 to 6.7 × 1.5	4.8 to 6.1 × 1.8
Capacity (t/h)	90–270	155–335	225–400	270–450	350–600
Power (kW)	10	15	20	25	35
Vibration (rev/min)	1400	1400	1400	1400	1400
Weight (kg)	3000	3500	4500	6000	8000

Table 28.8 'Belt feeders

Item	Belt width (m)					
	0.6	0.75	0.9	1.05	1.2	1.35
Belt speed (mm/s)	50–150	50–150	50–150	50–150	50–150	50–150
Capacity (t/h)	50–150	75–225	110–325	150–450	200–580	275–725
Power (kW)*	2–5	2–5	2–5	4–7.5	4–10	4–10
Weight (kg)	2500	2800	3500	4100	4900	5600

* Note (i) power depends upon length of belt (2.5 m min. to 6 m). (ii) A separate grizzle is needed, single or double deck. Dimensions 1.5 × 0.9 to 3.0 × 1.5 m are available. Power 4–15 kW.

SCREENING AND SCREENS

Once the material has passed through the primary crusher it together with the bypassed product is either temporarily stockpiled or directly transported to the next crushing phase. Before this takes place, however, particles smaller than the setting are screened out. Further separation into single sizes can also take place at this stage by using double or triple deck screens. These are commonly of the vibrating type or occasionally a revolving drum.

INCLINED VIBRATING SCREENS

The screen is generally made from interwoven wires to form a mesh with regular size openings through which smaller particles fall leaving the large fragments behind to be subsequently transferred to storage bins. Screens may

Figure 28.19 . Inclined vibrating screen

be arranged in tiers with each layer or deck having progressively smaller openings.

The whole unit is set on an incline with the material rolling off each deck being directed to separate storage bins. The screening motion is provided by a vibratory mechanism (fig. 28.20) in the form of a rotating heavy duty eccentric shaft.

Figure 28.20 Vibrating screen mechanism

HORIZONTAL VIBRATING SCREEN

These are similar to the above, but the decks are mounted horizontally and therefore afford considerable savings in head room. The vibratory mechanism is obtained by means of two counter rotating eccentric shafts, as described for the griddle. Thus the material is moved up and down and along the decks.

Figure 28.21 Horizontal vibrating screen

REVOLVING DRUM SCREEN

The drum rotates on a sloping axis, with material entering at the upper end and discharging progressively along the drum through perforations in the sides. The size of the perforations increase in stages along the drum length with the smaller material screened first into an outer drum as shown. The inclusion of a solid section extends the sieving time for the coarse material. Oversize material is simply allowed to escape at the lower end.

Figure 28.22
Revolving drum screen

CAPACITY OF SCREENS

Manufacturers usually state the quantity of material that can be passed through a single screen deck based on a standard aggregate, moisture content, etc. In practice, however, such conditions are unlikely to occur and therefore adjustments to these volumes are needed if a correct balance between crushing units, transporting systems and screening is to be maintained. This is best done by trial and error gained from experience with similar aggregates and screening units but approximate adjustment factors can be applied for shape and size of the aggregate, number of decks in the unit, moisture content, spraying, etc.

Deck Correction Factor

Not all the material which is of a size capable of passing through the screen openings actually does so and some will be retained, thus reducing the capacity (i.e. efficiency) of the screens. While this is normally assumed and included in manufacturer's output data for a single deck, when more decks are added the reduced efficiency of the system should be taken into account. Typical adjustment factors are:

No. of decks	Factor (x)
1	1.00
2	0.90
3	0.75
4	0.60

These values, however, are only a guide as the type and shape of aggregate will have an additional influence.

Size of Aggregate Factors

(i) Undersize aggregate: manufacturers usually base the screen capacity on the assumption that 45% of the aggregate falling on to the deck will be half the size of the openings in the deck. If this proportion increases then the screen will be able to handle more material and vice versa as shown below:

% Aggregate coming on to screen less than half screen opening size	Factor (y)
10	0.5
20	0.7
30	0.8
45	1.0
50	1.2
60	1.5
70	1.8
80	2.3
90	3.0

Again these factors should be tempered by other variables such as shape and type of aggregate.

(ii) Oversize aggregate factor: manufacturers also base their capacity figures on the assumption of 25–30% of the material coming onto a screen remaining there otherwise a small adjustment is required.

% Aggregate coming on to screen greater than screen size	Factor (z)
Less than 30	1.0
40	0.95
50	0.90
60	0.85
70	0.8
80	0.75
90	0.7

Shape of Aggregate, Spraying, etc.

Moist aggregate with fines reduces the efficiency compared to the dry condition, on the other hand where water is continuously sprayed on to the aggregate, as it is screened the rate of throughput can be increased. This technique is particularly suitable for medium fine mesh size, but is less effective for very fine and coarse material. When water is at a premium then the aggregate is best dried. This can be achieved with heated screens.

Screen opening size (mm)	Spraying factor (D) compared to dry conditions of screening
1	1.25
2	1.50
3	1.75
4	1.8
5	1.9
6	2.0
8	1.9
10	1.75
12	1.5
18	1.0

The capacity of the screen can be described as $W^1 = Q \times A \times X \times Y \times Z \times D$

Table 28.9 Capacity of Vibrating Screens
Capacity in tons per hour passing through 1 ft² (0.093 m²) or screen

	Size of clear square opening (in) (1 in = 25.4 mm)																							
	0.0117	0.0165	0.0234	0.0331	0.0469	0.0787	0.093	0.125	0.131	0.185	1/4	3/8	1/2	5/8	3/4	7/8	1	1 1/4	1 1/2	2	2 1/2	3	4	5
USS mesh size	50	40	30	20	16	10	8	7	6	4														
Sand	0.144	0.183	0.226	0.282	0.36	0.45	0.57	0.69	0.73	0.90	—	—	—	—	—	—	—	—	—	—	—	—	—	—
Stone dust	0.120	0.152	0.188	0.235	0.30	0.375	0.475	0.56	0.595	0.75	—	—	—	—	—	—	—	—	—	—	—	—	—	—
Coal dust	0.091	0.115	0.142	0.178	0.226	0.284	0.36	0.43	0.45	0.57	—	—	—	—	—	—	—	—	—	—	—	—	—	—
Gravel	—	—	—	—	—	—	—	—	—	—	1.08	1.40	1.68	1.94	2.16	2.36	2.56	2.90	3.20	3.70	4.05	4.30	4.65	4.90
Crushed stone	—	—	—	—	—	—	—	—	—	—	0.88	1.19	1.40	1.60	1.80	1.96	2.12	2.40	2.68	3.10	3.38	3.60	3.86	4.07
Coal	—	—	—	—	—	—	—	—	—	—	0.68	0.88	1.04	1.21	1.36	1.48	1.60	1.83	2.00	2.31	2.53	2.69	2.91	3.06

where

W^1	= volume per hour, e.g. m³ per hour;
Q	= manufacturer's capacity rating, e.g. m³ per hour per m²;
A	= Surface area of the screen, e.g. m²;
X, Y, Z, D	= adjustment factors, e.g. units.

Typical manufacturer's capacity values are shown in table 28.9.

Washing

Generally crushed aggregate can be adequately washed during the screening phase by applying water from sprays placed over the screen deck (fig. 28.23). Usually 50–150 litres per m³ of aggregate at 2–3 bar pressure is sufficient to wash away any clay or similar material adhering to the crushed fragments. Aggregate or sands (especially when obtained in the natural state from a pit) having more than about 3% by weight of clay or other contaminating material should be passed through separate washing equipment before screening. The choice of plant is dependent upon the size of the aggregate and amount of undesirable material present. The common methods are as follows:

(i) Simple wash drum.
(ii) Screw classifies.
(iii) Drum scrubbing.

Figure 28.23 Screen washing system

Simple wash drum

This is a rotating horizontal drum similar to the revolving drum used for screening (fig. 28.22), but with only a single coarse screen at the discharge end, the dirty water is simply led away or recycled after cleaning. The capacity depends upon the diameter and drum length but units capable of washing 200 m³/h of aggregate are available. The method is commonly used for producing clean washed chippings and grits. Typical data are given in table 28.10.

Screw classifier

The screw classifier is generally used for washing pit sands and comprises a feed box, settling hopper and rotating screw set at an incline. Water and sand are constantly added at the feed box with the dirty water overflowing to waste

Table 28.10 Rotary screen and wash drum

Item	Size of drum (dia. × length) (m)			
	0.8 × 3.8	1.0 × 4.2	1.2 × 5.0	1.5 × 5.8
Capacity (m³/h)	20–25	40–50	70–85	160–190
Water required (litres/s)	25	50	75	125
Power (kW)	5	8	12	20
Rotation (rev/min)	16	14	12	10

Figure 28.24 Screw classifier

or recycled. The sand particles are washed as they travel along the screw, emerging wet but suitable for transport away in trucks.

The rotation speed of the screw must be slow if the type of material to be washed is fine, otherwise particles passing the No. 50 sieve will be washed out.

Where screening is involved, the screw classifier is generally positioned to receive the screened fines and the screen wash water. Typical data are shown in table 28.11.

Scrubbing drum

For ores, stones and gravels and very dirty sands, washing generally requires a paddle drum. The equipment comprises a rotating drum fitted with paddles, spiral blades, etc. The aggregate enters at one end of the drum and leaves at the other, while water is introduced under pressure in the opposite direction. The whole unit can be positioned horizontally, but the retention time can be altered by providing an incline. Typical data for scrubbing washer drums are shown in table 28.12.

Figure 28.25 Scrubbing drum

Recovery of Fines from Wash Water

Fines capable of passing the No. 100 sieve will generally be lost in the wash water. However, when it is desirable to retain this material for subsequent

Table 28.11 Capacities in tph of screw classifiers

Material	Screw speed % of Max.	Size of classifier (screw dia. × length) (mm)							
		500 ×5750	600 × 6750	750 × 7500	900 × 7500	1000 × 8500	1200 × 9750	1300 × 10 500	1800 × 11 000
Very coarse sand	100%	30	45	70	100	140	185	290	420
Coarse sand	80%	25	35	55	80	110	150	230	335
Medium sand	60%	20	28	40	60	90	110	175	250
Fine sand	40%	12	18	27	40	55	75	115	170
Very fine sand	20%	6	10	13	20	30	40	60	85
Water (litres/s)									
100 mesh	—	35	42	48	60	95	120	130	150
200 mesh	—	10	12	14	24	30	35	38	

Table 28.12 Scrubbing drum data

Item	Size of drum (dia × length) (m)		
	1.8 × 3.2	*2.4 × 4.3*	*3.0 × 5.3*
Washing time (mins)	1–35	1–35	1–35
Capacity (t/h)	50–180	120–340	225–750
Water requirements (litres/s)	14–65	35–150	63–250
Power (kW)	60	150	380

Figure 28.26 Cyclone method of fines recovery

blending, fine particles in 30–200 mesh size can be recovered by a cyclone technique (fig. 28.26). The sand-bearing water is fed under pressure, tangentially into the top of a conically shaped drum. Near the walls centrifugal forces and boundary flow cause the sand particles to travel towards the outlet, while the secondary flows produce a circulating motion along the length of the cone together with a central vortex which is lead to an overflow pipe.

Typical Cyclone Data

Capacity	40–100 litres/s
Delivery	1–1.5 bar
Pump power	15–30 kW
Pump speed	650–850 rev/min

The alternative to the cyclone method is to use settling tanks or basins (fig. 28.27) but regular emptying to remove the sediment is needed.

Figure 28.27 Settling tanks for fines recovery

CONVEYOR SYSTEM

A belt conveyor is used to transport aggregate between crushing units, stock piles, screens and storage bins. The selections of capacities should be determined from a careful analysis of the whole system with imput to output related to usage rates of aggregate and sand. It is not enough to simply select the conveyor on the basis of the capacity of the crusher since individual units often carry smaller quantities to various screening and stock piles locations. Examples and data on conveyors are described in chapter 25.

Figure 28.28 Portable conveyor

EXAMPLE OF AN AGGREGATE PRODUCTION SOLUTION

Problem

Quarry rocks of 350 mm size are delivered to a crushing plant at a rate of 50 t/h. The sizes required are

$1\frac{1}{2}$ in (38 mm) to $\frac{3}{4}$ (19 mm)
$\frac{3}{4}$ in (19 mm) to fines.

Select appropriate crushers and screens.

Solution

PRIMARY CRUSHING

Maximum allowable size of material is 0.8 × feed opening. Thus minimum feed opening required = 350/0.8 = 438 mm.

Figure 28.29 Crushing and screening example

225

Alternative options

Interpolating from table 28.2 for the 0.45 × 0.8 m jaw crusher:

Closed setting = 50 mm   output = 52 t/h
Closed setting = 180 mm   output = 130 t/h

From table 28.3 for 0.33 × 1.1 gyratory crusher:

Open setting = 50 mm   output = 80 t/h
Open setting = 75 mm   output = 150 t/h

Thus the jaw crusher is selected with a 50 mm setting to produce the required output of 50 t/h.

From fig. 28.30 illustrating an analysis of product from jaw crushers it can be seen that for a 50 mm setting 63% of material passes the $1\frac{1}{2}$ in (38 mm) screen and can be further screened to produce the $1\frac{1}{2}$ in (38 mm) to $\frac{3}{4}$ in (19 mm) and $\frac{3}{4}$ in (19 mm) to $\frac{3}{16}$ in (5 mm) size products. The 37% of oversize material or 18.5 t/h is directed to a secondary crusher.

Figure 28.30 Jaw crusher product sizing graph

SECONDARY CRUSHING

Typically the cone crusher could be selected as the secondary crusher. It can be seen in fig. 28.29 that the oversize 18.5 t/h consists of pieces up to 76 mm size. Thus the feed opening of the crusher should be approximately 75/0.8 = 94 mm.

Table 28.5 which gives typical cone crusher data indicates that a unit with a 600 mm diameter cone is adequate as follows:

Feed opening   90 mm
Discharge setting   6–25 mm
Capacity   13–50 t/h

Figure 28.30 shows that with a $\frac{3}{4}$ in (19 mm) setting all the stone would pass the $1\frac{1}{2}$ in (38 mm) screen and so can be screened into the two required grades of material.

SCREEN SIZING

Screening after Primary Crusher
$$\text{Screen capacity} = Q \times X \times Y \times Z \times D$$

Top deck (38 mm screen)

From Table 28.9 for crushed stone passing the 38 mm ($1\frac{1}{2}$ in) screen

Q = 2.68 t/h per ft^2 (29 t/h per m^2).

Factor X: for top deck = 1.00.

Factor Z: 37% of material to be screened is larger than 38 mm, i.e. 37% oversize. Factor = 0.965.

Factor Y: from fig. 28.30:
30% of the total material coming onto the screen is less than half screen size (19 mm).
Therefore factor = 0.8.

Factor D: = 1.0.

If 50 t/h are delivered then the required top screen area is

$$\frac{50}{29 \times 1 \times 0.8 \times 0.965} = 2.23 \text{ m}^2$$

Lower deck (19 mm screen)

Factor X: for lower deck 0.9.

Factor Z: from fig. 28.30 30% of 50 tonnes passes the 19 mm sieve, i.e. 15 tonnes; 63% passed the 38 mm sieve, i.e. 31.5 tonnes, therefore 16.5 tonnes remain on the 19 mm sieve i.e. 16.5/33.5 = 49%.
Factor = 0.91.

Factor Y: from fig. 28.30 15% of the total material coming onto the screen is less than half screen size (9.5 mm).
Factor = 0.6.

Factor D: = 1.0.
From table 28.9 for crushed stone passing the 19 mm ($\frac{3}{4}$ in) screen:

Factor Q: = 1.80 t/h per ft^2 (19.35 t/h per m^2).

If 31.5 t/h pass the top deck then the required bottom deck screen size is

$$\frac{31.5}{19.35 \times 0.9 \times 0.6 \times 0.91 \times 1.0} = 3.3 \text{ m}^2$$

Commonly a twin-deck vibrating screen would be chosen and in this case the dimensions would be dictated by the bottom screen requirement. Thus 1.2 × 3.0 m screens would be adequate.

Screening after Secondary Crushing (19 mm screen)
18.5 t/h is passed through the secondary crusher which has a 19 mm ($\frac{3}{4}$ in) setting.

Figure 28.31 Cone crusher product sizing graph

CURVE No	SETTING	
1	⅛ in	3.2 mm
2	¼ in	6.5
3	⅜ in	10
4	½ in	12.5
5	¾ in	19
6	1 in	25
7	1¼ in	32
8	1½ in	38
9	2 in	50

Factor Q: = 1.8 t/ per ft² (19.35 t/ per m²).

Factor X: = Single deck = 1.0.

Factor Y: From fig. 28.31 35% of the total material coming on to the 19 mm screen is less than half screen size. Factor = 0.86.

Factor Z: 33% (6.105 t/h) of material to be screened is larger than 19 mm. Factor = 0.98.

Factor D: 1.0.

If 18.5 t/h is passed then the required screen area =

$$\frac{18.5}{19.35 \times 1.0 \times 0.86 \times 0.98 \times 1.0} = 1.13 \text{ m}^2$$

PORTABLE SCREENING EQUIPMENT

Fixed plant is usually set up where crushed stone is required over a long period and the transporting costs to the customers are economical. However, the many instances where rock is available for crushing in the nearby vicinity

Figure 28.32 Portable crushing and screening equipment

has led to the development of portable equipment. The early developments were fully self-contained portable units comprising crushing, conveying and screening. More modern plant, however, has concentrated on providing individual processing stations mounted on their own chassis. Thus a single station might comprise the screens or feeder and crushing machine. Portable conveyors are used to connect individual units and storage bins.

Chapter 29   CONCRETE PRODUCTION

CONCRETE DESIGN

Concrete consists of a mixture of graded sand, graded coarse aggregate, Portland cement and water. By itself cement mixed with water causes a chemical reaction producing a hardened paste after about 45 min, called the 'initial set'. At not more than 10 hours this plastic phase is over and the final set obtained. Provided sufficient water is present for the reaction to continue, the compound will gain further strength until about 28 days when almost the final strength is reached, the rate of gain depending upon the surrounding temperature.

The addition of aggregate particles simply serves to act as filler material bound together by the cement paste. Thus the strength and grading of these materials are influential in controlling the strength of the hardened concrete.

CEMENTS

Portland cement klinker is commonly manufactured in a rotating kiln by gradually heating and finally burning a finely ground mixture of calcium carbonate (chalk, limestone) and siliceous material (clays). Different types of Portland cement can be produced with various properties by modifying the chemical composition and final grinding of the klinker to conform to particular National Standards including specification for fineness, strength, setting time and limits of impurities. The more common Portland cements are:

(i) Ordinary Portland Cement (OPC) – a basic cement used for general concrete. Fineness: 30 mm²/g.
(ii) Rapid hardening Portland – a finer cement used to give high early strength.
(iii) Low heat Portland cement – used for massive concrete pours such as dams, to reduce the heat of hydration generated during the chemical reactions.
(iv) Sulphate-resisting Portland cement – made by adding iron compounds during the kiln process to produce a material less affectable to acid waters and other injurious salts.

(v) Other cement varieties include, coloured, blast furnace, pozzolanic, masonry, waterproof, hydrophobic, high alumina and oil well cements.

(vi) Pulverised fuel ash may be combined with OPC to produce structural concrete. The resulting mix gives a slightly increased workability.

AGGREGATES

Aggregates occupy about 75% of the volume of concrete and are obtained either from naturally occurring deposits of sands and gravels, including sea-dredged material, or are produced by crushing quarried rock. Granite, basalt and the harder types of limestone and sandstone are commonly used. They should be free of impurities, clay, silt and degradable materials. A simple quality test for sand can be carried out by filling a 200 ml measuring cylinder up to the 100 ml mark and then adding water up to 150 ml mark. After shaking the mixture and allowing about 3 hours for settling, the thin layer of silt formed should not exceed 10% of the sand volume.

GRADING OF AGGREGATES

To obtain good strength, concrete should contain particles ranging in size to facilitate an interlocking of the stones. In achieving this the aggregates can be obtained in single sizes and then combined to produce suitable gradings but are more commonly supplied already graded. Fine aggregate or sand is that material passing through a 5 mm ($\frac{3}{16}$ in) sieve, while coarse aggregate is retained on this size of sieve, (Table 29.1). The maximum size of coarse depends upon the job requirements, for example reinforcement congestion would dictate small size aggregate, whereas mass concreting might be suitable with 75 mm or more stone size.

CONCRETE MIXES

WATER/CEMENT RATIO

While the quality and grading of the aggregate are important, the strength of concrete depends mainly on the water/cement ratio and the type of cement, the water cement ratio being defined as the weight of water in a mix divided by the weight of cement. The higher the w/c ratio, the weaker the concrete (fig. 29.1) becoming significantly pronounced at about 0.6 w/c and above. Indeed, a water content of only 25% of the weight of cement is needed to combine chemically, the higher amounts being required to provide workability. High water/cement ratios also increase permeability.

DEGREE OF COMPACTION

The strengths shown in fig. 29.1 assume a fully compacted concrete. Unfortunately during mixing air is entrapped and unless sufficient vibration is

231

Table 29.1 Coarse and fine aggregates

Sieve size (mm)	Percentage passing							Fine aggregate
	Graded aggregate			Single-sized aggregate				overall limits
	40 to 5 mm	20 to 5 mm	14 to 5 mm	40 mm	20 mm	14 mm	10 mm	
50	100	—	—	100	—	—	—	
37.5	90–100	100	—	85–100	100	—	—	
20.0	35–70	90–100	100	0–25	85–100	100	—	
10.0	10–40	30–60	50–85	0–5	0–25	0–50	85–100	
5.00	0–5	0–10	0–10	—	0–5	0–10	0–25	100
2.36	—	—	—	—	—	—	0–5	90–100
1.18								30–100
600 μm								15–100
300 μm								5–70
150 μm								0–15

Figure 29.1
Compressive strength and water/cement ratio

Figure 29.2
Compressive strength and voids ratio

provided the ultimate strength will be seriously affected as illustrated in fig. 29.2.

WORKABILITY

Workability describes the degree of stiffness of the mix and can be directly altered by changing the water/cement ratio. However, the shape, size and texture of the aggregate also have some effect. For example sea-dredged rounded stones generally produce more workable mixes than crushed angular material: also less water is required for a given workability by increasing the maximum size of stone.

Tests for workability

(i) *Slump test* (fig. 29.3). A conical mould is filled in three equal layers and rodded twenty-five times per layer. The top is struck level, the mold removed and the slump measured. Low workability is usually indicated by a 25 mm slump and less, medium by 25–75 mm, and high greater than 75 mm. A collapsed or sheared slump should be repeated.

Figure 29.3 Slump test

Figure 29.4 Compaction factor test

(ii) *Compaction factor test* (fig. 29.4). Concrete is dropped from vessel (1), into (2) and finally into cylinder (3). The final surface is smoothed level and the cylinder and contents weighed. The cylinder is then refilled and rammed to full compaction.

$$CF = \frac{\text{weight of partially compacted concrete}}{\text{weight of fully compacted concrete}}$$

(iii) *Vebe test* (fig. 29.5) is used for very stiff concrete mixes and consists of vibrating a standard slump test specimen placed in a specified vessel. The workability is expressed in seconds of time required to level the cone, e.g. 10 mm slump is approximately equivalent to 12 s, 30 mm ≡ 6 s, 60 mm ≡ 3 s, 180 mm less than 1 s.

(iv) *DIN test* (fig. 29.6) is designed for very high workability mixes, i.e. those almost of flowing consistency. A shortened conical mould is filled in two layers and rodded ten times. The mould is then removed and the foundation board lifted, tilted and dropped fifty times in 15 s. The average spread of concrete denotes the workability measure.

PROPORTIONING OF CONSTITUENT MATERIALS

The cement, water and aggregates are usually defined in terms of kilograms of each material per cum of concrete, determined by one of various well established standard methodologies, e.g. BRE method in the UK supported with trial mixes. Typically: 250 kg/m^3 cement, 125 kg/m^3 water, 1500 kg/m^3 aggregate.

Figure 29.5 Vebe test

Figure 29.6 DIN test

Determining the ideal combination of coarse and fine aggregate in particular is best achieved by trial and error, making reference to appearance, workability and cement content in achieving the required 28-day strength. Alternatively proportions may be roughly gauged from specified grading curves, charts, tables, etc.

MOISTURE CONTENT OF THE AGGREGATE

Generally aggregates when arriving at site are wet from the washing operation. Further changes in the moisture content are likely to take place during storage out in the open atmosphere. The weight of this water should be deducted when measuring out the quantities during batching and may be ascertained from weighing wet and dried samples or by proprietary chemical methods. Unfortunately, some of the water may be absorbed as pore water in the aggregate and while contributing to the water/cement ratio would tend to produce less workability than expected.

Example of Theoretical Adjustments

Consider a 1 cum batch of concrete

Sand	Aggte.	Cement	Water
600 kg	1200 kg	300 kg	150 kg

Moisture content for sand = 5%

$$\text{weight of water} = \frac{600 \times 5}{100} = 30 \text{ kg}$$

Moisture content for coarse aggregate = 3%

$$\text{weight of water} = \frac{1200 \times 3}{100} = 36 \text{ kg}$$

Batching weights (approximately)

Sand	Aggte.	Cement	Water
630	1236	300 kg	84

Mixing water should be free from impurities such as organic materials and injurious chemicals. Tests for water for making concrete are laid down in various National Standards.

CONCRETE STRENGTH

SETTING TIME

Concrete, unlike cement paste, is not described in terms of initial and final setting times, the set being more practically determined according to the time trowelling or re-vibrating cease to be possible. This condition is reached after about ½ hour in warm air temperatures and 4 hours or so when cold. The inclusion of a retarding admixture can further extend this period.

STRENGTH GAIN

Strength continues to improve provided sufficient water is supplied to the set concrete to ensure that drying out during the heat of hydration stage is prevented, otherwise the chemical reactions which allow strength to be developed, will cease. The process is known as 'curing'.

Shrinkage occurs as concrete hardens, being greater in wet mixes. Curing helps to limit its effects and therefore concrete in the actual structure should be thoroughly and continuously moistened for some considerable period after placing.

Figure 29.7
Compressive strength
gain with age

TEST SPECIMENS

Figure 29.8 Typical
crushed test cube

Concrete strength is usually measured according to various National Standards by carrying out a compression test on samples taken from the mix. A cube or cylinder specimen is crushed to destruction (fig. 29.8) and the sustained maximum load per unit area measured. The 28-day value is normally selected to represent the characteristic compressive strength. Depending upon the required concrete properties grades from 2.5 to 60 N/mm^2 can be achieved for normal mixes, but with suitable admixtures and strong aggregates up to 200 N/mm^2 is possible. Unfortunately, 28 days is often too long to wait and methods of accelerated curing of test specimens have been developed. They principally involve placing the test specimens and mould into a tank of heated water shortly after making and leaving it for a period before crushing, normally 24 hours or so depending upon the temperature. Typically accelerated tests have been developed for 35 °C, 55 °C and 82 °C whereby the early and 28-day results are correlated and thereafter used as a gauge for all further testing of specimens.

Test specimens may also be obtained as core samples taken for the actual concrete structure or component or alternatively by non-destructive means such as ultrasonic, gamma-ray, etc. Other surface investigation methods include the Schmidt hammer, internal fracture and pull out tests.

Interpreting Test Results

When a large number of test samples are crushed to destruction, the usual pattern of strength is a normal distribution (fig. 29.9). To ensure that no

Figure 29.9 Normal distribution of crushed test cube results

results fall below, say, a permitted minimum crushing strength the average strength of the mix would need to be considerably raised. In practice however a small number of results are allowed to fall below the minimum, calculated from the following formula:

Minimum crushing strength = average crushing strength $-2 \times$ standard deviation that is $x = \bar{c} \pm Z s$

where

$$s = \text{standard deviation} = \left[\frac{\Sigma \, (c - \bar{c})^2}{n-1} \right]^{\frac{1}{2}}$$

and $\bar{c}$ = average strength of all test results;

c = each individual test result of strength;

n = number of results;

Z = standard normal variable, obtained from statistical tables, e.g. 1.96 for 95% confidence level.

Example

A large number of test results produce a standard deviation (s) 1.5 N/mm^2, and average strength 10 N/mm^2. If 5% of results are permitted to fail, what should be the minimum design strength of the mix.

$$x = \bar{c} - Zs$$
$$= 10 - 1.96 \times 1.5 = 7.06 \, \text{N/mm}^2$$

That is, if the specified minimum strength were 7 N/mm^2 then a mix with an average strength of nearly 10 N/mm^2 would be necessary if not more than 5% of test results are not to fall below the specified strength.

ADMIXTURES

Chemicals are available for addition to the mix to act as set retarders, set accelerators, strength accelerators, water reducers, air-entraining agents, water repellents, thickening agents, superplastisicers, pigmenters. The amounts and use of such additives vary depending upon the application and manufacturer.

OTHER TYPES OF CONCRETE

Concrete can be made from various materials to suit specific needs. The

Table 29.2 Typical uses of some admixtures

Use	Set retarder	Set accelerator	Strength accelerator	Water reducer	Air-entraining agent	Water repellent	Thickening agent	Superplasticiser	Pigment
Chemical resistance				*	*				
Pumping	*			*	*		*	*	
Early formwork removal			*						
Workability aid				*	*			*	
High early strength		*	*	*				*	
Permeability reduction				*	*	*			
Mortars					*			*	*
Precast cladding			*		*	*			*
Frost resistance				*	*				
Mass construction	*							*	
Waterproofing				*	*	*			
Improved surface finishes					*	*		*	
High strength				*				*	
Hot weather	*			*					
Cold weather		*	*						

following are typical examples: clinker, no-fines, light-weight aggregate concrete, high strength, dense aggregate concrete.

CONCRETE REINFORCEMENT

Cutting and bending of reinforcement can be carried out on site or ready supplied. The steel bars are fixed together by tie wire or welding. Mats of steelwork are supported by 'stools' or 'chairs'. Individual bars may be given sufficient cover of concrete by providing plastic or simple concrete spacer blocks attached to the reinforcement at points facing the formwork. In normal practice mild steel is usually specified but occasionally high tensile steel may be stipulated by the designer.

DEFECTS IN CONCRETE

BLEEDING

When the solids settle under the action of gravity bleeding results to leave a layer of water, and consequently a porous and weak final surface. Increased cement and/or the use of an air-entraining agent in the mix will help to counter the effect.

SEGREGATION

When the cement paste separates from the aggregate segregation arises and is generally caused or aggravated by rough handling during the placing operation. The problem is particularly acute with excessively sanded mixes, when the coarse aggregate tends to separate out to the bottom during vibration.

SULPHATES

Concrete exposed to sulphates, e.g. sea water, contaminated ground, etc., will undergo chemical change, expand and may crack, the process being mainly caused by the calcium hydroxide and tricalcium aluminate hydrate in the cement reacting with sulphate ions to produce solid products greater in volume than the original.

ACIDS

Any acidic liquid in contact with concrete will produce some dissolution, the degree of harm being dependent upon the type and concentration of acid present. In general, however, acid attack in most grounds is not a major problem.

SALTS

Dissolved salts can penetrate concrete, any subsequent crystalline growth may then exert considerable forces and cause damage.

CARBONATION

The hydroxides in concrete may become neutralised near the surface by the action of carbon dioxide in the atmosphere, and destroy the alkaline protection to the embedded reinforcement. A potential difference could then be established between two such regions and metal ions transferred. Any water and oxygen present then causes a reaction to produce ferrous hydroxide, i.e. corrosion in the carbonised layer. The depth of cover should thus be sufficiently deep to reduce such an effect.

CHLORIDES

The presence of chlorides in the concrete can produce an electrolytic solution and thereby induce corrosion in reinforcement by causing local potential drop in adjacent parts of any embedded steel where chlorides are present. The maximum total chloride content of aggregates, and cement are laid down in National Standards but should not exceed 0.3% by weight of cement, the only reliable means of reducing the problem is to provide cathodic protection to the steel. Sea-washed aggregate should also be carefully scrutinised for chloride content.

ALKALI–SILICA REACTION

Most cement contains small amounts of sodium and potassium alkalis, e.g., NA_2O, K_2O, which in the presence of water and some forms of silica aggregate, for example, opal reacts to produce a gel and subsequent disruptive expansion and associated cracking. A low alkali cement containing less than 0.6% by weight of Na_2O, etc., is recommended to reduce or at least limit the effect.

PRACTICAL CONSIDERATIONS

WATER CONTROL

1. Cover all exposed concrete surfaces with damp sand, hessian, etc., and keep moist.
2. Maintain formwork in place for as long as practically possible within the work programme (the minimum period is often specified in the contract and will generally vary according to the nominal grade of concrete, and whether the structure is columns, walls, slabs, foundations, etc.). Thereafter drape damp hessian over the exposed surfaces and keep wet.
3. Alternatively, spray any exposed concrete surface with a waterproof membrane to help hinder the drying out process. Thereafter keep the concrete continuously wet.
4. Curing should always continue for from 3 to 7 days after formwork removal according to the prevailing temperature conditions. In any case a strength of at least 2 N/mm^2 is needed to resist freezing forces and 10 N/mm^2 to support temporary working loads.

TEMPERATURE CONTROL

Cold Conditions

First and foremost fresh concrete should be prevented from freezing by applying one or more of the following:

1. Heat the water and aggregates prior to mixing. The latter can be achieved with steam pipes placed in the stockpiles.
2. Insulate and heat the formwork, usually obtained with electric cabling.

3. Heat the surrounding air in enclosed situations.
4. Transport concrete in insulated or heated vehicles and containers.
5. Introduce suitable admixtures (see table 29.2).

Hot Conditions

In hot climates evaporation needs to be controlled to avoid rapid stiffening, shrinkage and plastic cracking. The following preconditions should be taken:

1. Insulate formwork.
2. Provide shading to reflect sun rays and cover all exposed surfaces with wet hessian, etc.
3. Fog spray.
4. Cool the concrete with embedded pipework. The heat extracted can be theoretically determined by the formula:

$$N = Q\gamma_w S_w t$$

where N = heat quantity extraction rate, e.g. kcal/hr;
  Q = cooling fluid passed, e.g. m^3/h;
  γ_w = specific weight of cooling fluid, e.g. water = 1000 kg/m^3;
  S_w = specific heat of cooling fluid, e.g. water = 1 k cal/kg °C;
  t = exit and entry temperature difference of cooling liquid (°C). Initially the exit temperature will be that of the fresh concrete and should not exceed 30–35 °C (15 °C for mass concrete).

Note: total heat to be extracted = $WS_c t$
where W = weight of concrete
  S_c = specific heat of concrete.

5. Introduce blocks of ice, solid CO_2 or liquid nitrogen into the mix and cool the aggregate; the temperature of the mix should be kept below 35 °C or so.
6. Introduce suitable admixtures (see table 29.2).

CONCRETE BATCHING AND MIXING

BATCHING

A hopper is filled with the specified weight of each material obtained from the storage bins and the contents transferred to the mixing drum. Cement is usually pumped from a silo, weighed and discharged separately, similarly with admixtures. When all the ingredients are in the mixing drum the required quantity of water is added from a water-measuring tank, water meter, etc. Depending upon the sophistication of the equipment the process may be manual with the operator having to watch the scale dials carefully, or alternatively on some modern central mixing units automatic gate controls close as the specified weight of material is loaded. Regular servicing and checking of the weighing mechanisms is crucially important to maintain batching accuracy.

MIXING

The sequence of discharging materials and water to the mixing drum varies depending on the type of equipment and is best determined by trial and error judged by the uniformity of the concrete produced. In modern plants roughly 1 min is required for charging the drum and a further 1 to $1\frac{1}{2}$ min needed for mixing. The choice of mixer being either drum or pan types.

Drum Mixers

These comprise a bladed vessel, usually of conical shape having one opening both to receive and discharge the ingredients, rotating on an inclinable axis (fig. 29.10) to facilitate discharge away from the loading point. Small to large units are available capable of producing mixed batches in the range from 5 to 100 ft^3 (140–2800 litres), with outputs of from 5 to 120 yd^3/h (4–90 m^3/h). The mixing action is produced by helical baffles moving the material from the charging end to the base of the drum. Reversing the rotation causes a counter reaction to discharge the contents. Rotation speeds of up to 10 rev/min are normal. An alternative form of this type of mixer is *non-tilting* with the drum rotating on a horizontal axia and having two openings (fig. 29.11) one to receive materials, the other to discharge the mix. Reversing action is not required and as a consequence are normally of simpler design, having a relatively lightweight support frame. However, the discharge time is longer than for the tilting type. The size of vessel should be capable of holding as least 10% more than its nominal capacity. The drum volume itself is often up to four times the nominal capacity to allow sufficient space for the mixing action.

Figure 29.10 Tilting drum mixer

Figure 29.11 Non-tilting drum mixer

241

Pan mixers

The mixing vessel is mounted on a vertical axis (fig. 29.12) with the contents entering at the top and discharging through a sliding trap in the base and down a chute into transportation vehicles. Mixing is achieved with blades rotated within the vessel at 14–30 rev/min which on some models is also rotated counter to the blades. Pan sizes vary from about 5–90 ft^3 nominal capacity (140–2500 litres) capable of producing 5–130 yd^3/h (4–100 m^3/h).

Figure 29.12 Pan mixer

CENTRAL MIXING PLANTS

When large quantities of concrete are required, a central mixing facility with well prepared storage bins, and semipermanent equipment is usually installed. Good access points for distributing materials and transportation of concrete are essential.

Figure 29.13 Central mixing plant

HANDLING AND TRANSPORTING CONCRETE

WHEEL BARROWS

These are one or two wheeled pneumatic tyred barrows of capacity from 0.1 m^3 to 0.2 m^3.

1. Output per man on a 50 m return run is $1 \text{ m}^3/\text{h}$. Gang strengths of up to six men are common.
2. When using the two wheel type, output may increase to $1\frac{1}{2} \text{ m}^3/\text{h}$ but due to the increased weight, difficulty in turning and transporting up slopes, efficiency is generally reduced.

Uses

Chimneys, cooling towers, floor construction at ground level or in multistorey construction, confined areas.

HOISTING WITH A CRANE (FIGS. 13.5, 14.6, 16.1, 19.3)

Three types of crane are used:

1. Crawler-mounted – used for general all-purpose work such as pouring concrete in foundations, retaining walls, floor slabs, culverts.
2. Derricks – used for pouring concrete in pump houses, caissons, and basement construction where erection costs can be justified, such as work continuing for long periods.
3. Tower cranes – used for concreting floors, columns, multistorey construction, ground level work where the area to be covered is large.

Buckets and Skips

These range from $\frac{1}{2}$ to 8 m^3 capacity, the common types being $\frac{1}{2}$ and 1 m^3 skips. The bottom of the skip is fitted with a gate to allow the discharge to be manually controlled. The skip is hung from a crane.

Outputs using cranes

Ground level work	$\frac{1}{2} \text{ m}^3$ skip	8 to $12 \text{ m}^3/\text{h}$
	1 m^3 skip	14 to $20 \text{ m}^3/\text{h}$
Above ground level	$\frac{1}{2} \text{ m}^3$ skip	5 to $8 \text{ m}^3/\text{h}$
	1 m^3 skip	10 to $14 \text{ m}^3/\text{h}$

CONCRETE PUMPS

Hydraulic Pump

Concrete is placed in the pump hopper and forced along a 100–150 mm diameter steel pipe to the point of discharge.

Uses – large foundations, multistorey construction, floors slabs.

Outputs – (a) single cylinder, from 12 to $15 \text{ m}^3/\text{h}$; (b) double cylinder, from 15 to $50 \text{ m}^3/\text{h}$.

Typical range – 400 m horizontally or 35 m vertically.

Losses – A 45° bend loses 6 m horizontally.

Note: Small diameter (50 mm) pipe is now on the market. Such pumps are commonly mounted on trucks and are hydraulically operated. Generally for pumping purposes concrete should have a slump of about 75 mm. Pumping can be improved with lubricating additives (retarders) and PFA.

Pneumatic Pump

This consists of a cylinder containing the concrete. Compressed air is then introduced forcing the concrete to the place of discharge.

BELT CONVEYORS

These can now offer comparable costs with other forms of placing method. The concrete is simply transported from the mixer to the discharge point by the conveyor belt.

Output – depends upon the speed and size of the belt, transportable unit now on the market gives from 7 to 16 m³/h.

DUMPER

Used for transporting concrete from the mixer to the works and have capacities ranging upwards from 1 m³.

TRANSIT MIXERS (TRUCKS)

Mobile trucks (fig. 29.14) are very popular for delivering ready-mixed concrete to site and are usually of the inclined-axis revolving-drum type, discharging by a counter rotating action. Mixed concrete may be loaded at a central batching plant or mixing undertaken while travelling. Nominal capacities are available from approximately 2–6 m³ depending upon the permissible axle load and weight of truck allowed on the highway.

Figure 29.14 Ready-mixed concrete truck

MONO-RAILS

(fig. 26.1) – consist of concrete hoppers mounted upon a single rail. A series of these can be used to deliver concrete very efficiently on many types of works such as the concreting of bridge decks or similar where conventional access might be difficult.

PLACING CONCRETE

1. Concrete should be placed within 30 min of mixing.
2. Forms should be blown clean of debris and oiled to aid subsequent removal before concreting operations begin.
3. Concrete should be spread evenly in 300 mm layers to avoid segregation.
4. Concrete should not be dropped more than 2 m vertically and a chute used wherever possible to avoid segregation (fig. 29.15).

Figure 29.15 Placing concrete from a skip

5. Finishing of concrete surfaces may be achieved by manual or power floating methods after the surface has been tamped level and fines brought to the top. In cold weather up to 5 h may be necessary for sufficient hardening for the smoothing process to begin.

Tremmie tubes – where the height of placing concrete exceeds 2 m, a tremmie pipe should be used, (fig. 29.16).

Figure 29.16 Placing concrete with a tremmie tube

CONCRETING UNDER WATER

Either use a tremmie or a bottom-opening skip. The concrete should be carefully placed otherwise the fines will be washed out to form a layer of latence on top of the concrete. Even so, in large pours this layer may be up to 400 m deep. A high slump concrete (150 mm), and about 25% more cement giving a water/cement ratio of about 0.5 may be necessary.

COMPACTION

Concrete must be consolidated to remove voids which seriously reduce the strength of the hardened concrete, e.g. by up to 30% with 5% voids and 25% voids the concrete might only have 10% of the expected strength of the fully compacted value.

Compaction may be achieved with internal vibrating pokers inserted at $\frac{1}{2}$ m points as the concrete is placed, and slowly withdrawn when air bubbles cease to come to the surface. External vibrators clamped to the outside of the shuttering are an alternative system.

CONSTRUCTION JOINTS

Such joints are required to combat shrinkage, contraction and expansion and are normally required as follows:

Vertical joints – concrete floors and ground slabs are normally broken up by construction joints into bays or strips about 6 or 7 m wide and up to 20 m long. Walls are normally divided into panels with vertical construction joints at about 10 m centres.

Horizontal joints – are normally placed in walls and columns at 2 to 3 m centres vertically.

All joints should be formed level and thoroughly cleaned before the next concrete is placed.

WET CONCRETE PRESSURE

Fresh concrete placed in formwork for slabs (fig. 29.17), beams (fig. 29.18), walls (fig. 29.19) and columns (fig. 29.20) will exert a pressure on the sides and base until set. The formwork thereafter is left in place until sufficient strength has been obtained in accordance with the specification laid down by the designer of the structure.

Figure 29.17 Soffit formwork to a concrete slab

(a) PROPRIETARY SYSTEM

(b) TIMBER

Figure 29.18 Beam
formwork

Traditional in Timber

Proprietary

(a) WALL FORM

SUPPORT
FOR 1st
POUR ONLY

(b) CLIMBING WALL FORM

Figure 29.19 Wall
formwork

During the setting period the concrete will behave similar to that of a liquid contained in a vessel and thus the formwork needs to be sufficiently strong to both resist the 'fluid' forces and limit deflections to reduce the possibility of leaks. Facing formwork is usually made of timber boards or plywood sheeting suitably framed with bracing, all assembled into practical sized units to suit the method of site handling, erecting and removal. Commonly a proprietary system made up of steel panels may be chosen when many uses are required as plywood normally deteriorates rapidly after six or so uses. The whole system is assembled manually and opposing faces cross tied (fig. 29.21). In designing the formwork the weight of the ancillary items all may need to be taken into account as additions to the forces exerted by the wet concrete itself. Typical values are:

Figure 29.20 Column formwork

Figure 29.21
Formwork cross-ties

Self-weight of concrete	$25 \, \text{kN/m}^3$
Steel reinforcement	$80 \, \text{kN/m}^3$
Solid softwood timber	$5.5 \, \text{kN/m}^3$
Solid hardwood timber	up to $10 \, \text{kN/m}^3$
Proprietary panel formwork	from 0.5 to $0.85 \, \text{kN/m}^2$
Timber formwork	$0.5 \, \text{kN/m}^2$
19 mm thick plywood	$0.11 \, \text{kN/m}^2$
Superimposed load of workmen	$1.5 \, \text{kN/m}^2$
Superimposed load of equipment	$7 \, \text{kN/m}^2$
Horizontal impact loads	10% of total static load
Winds loads	varies

PRESSURE FORMULAE

The maximum pressure, (P_{max}), would not be greater than the total head, H, of concrete (excepting above factors), i.e

P_{max} not greater than γH

where γ = concrete density at $25\ kN/m^2$ depending upon mix properties.

In practice P_{max} may be reduced depending upon the rate of pour, temperature, vibration, weight of concrete, slump and arching, subject to the formulae used.

FACTORS AFFECTING CONCRETE PRESSURE

Rate of Placement
The setting of the concrete varies with time, therefore the faster the rate of pour the greater the horizontal pressure exerted on the face of formwork. Conversely, if the rate were slow enough the pressure at the bottom of the formwork would not be greater than at the top.

Effect of Temperature
Concrete at low temperatures sets more slowly than at high temperatures. Indeed for a given rate of pour the pressure exerted can be from 50 to 75% higher in cold weather compared to that under warm conditions.

Effect of Workability
The stiffer the concrete the better the ability to stand unsupported, thereby reducing pressure – a very workable mix behaves like a liquid.

Effect of Arching
Concrete placed between narrow gaps demonstrates similar characteristics to other granular materials and some arching will develop. The effect being to reduce pressure.

The combination of all these influences is to alter the normal triangular hydrostatic pressure design, with the maximum value shifted away from the base (fig. 29.22).

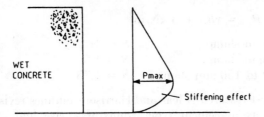

Figure 29.22 Concrete pressure diagram

USA FORMULAE

$$P = 150 + \frac{9000R}{T},\text{where } R \leq 7 \text{ ft/h}$$

$$P = 150 + \frac{43\,400}{T} + \frac{2800R}{T} \text{ ,where } R > 7 \text{ ft/h}$$

where P = maximum lateral pressure (in lb/ft^2);
       R = rate of placement (in ft/h);
       T = temperature of concrete in form (in °F);
       H = maximum height of concrete in the form (in ft).

OTHER FORMULAE

In the UK the effect of arching (fig. 29.23) is also included in some formulae, the pressure selected being the minimum value obtained from the following:

Figure 29.23 Arching of wet concrete

(i) Pressure head obtained by depth.
$$P = \gamma H \text{ kN/m}^2$$
where H = height of wet concrete (in m)
      γ = density of wet concrete (in kN/m^3)

(ii) Arching effect.
$$P = 3R + \frac{d}{10} + 15 \text{ kN/m}^2$$

where d is the width of the section (in mm). The formula only being valid for $d < 500$ mm. R is the rate of placement in m/hr.

(iii) Stiffness effect. S varies with both slump and temperature, as determined from tables. Typically for 15 °C concrete:

$$P = \gamma RS + 5 \text{ kN/m}^2$$

50 mm slump, $S = 1$;
75 mm slump, $S = 1.5$;
100 to 150 mm slump $S = 1.75$.

The latest research (Clear and Harrison) outlines revised formulae to include the effects of cement type, admixtures, size and shape of formwork. In general, however, the results are not dissimilar to other formulae.

EXAMPLE

Calculate the maximum pressure produced when placing 75 mm slump concrete at a rate of 2 m/h, shuttering 2.5 m high and 250 mm wide. Assume the temperature of the concrete in the formwork is initially 15 °C and vibration is continuous.

Solution
UK formulae

(i) By height $\quad P = 25 \times 2.5 = 62.5 \text{ kN/m}^2$

(ii) By arching $\quad P = 3 \times 2 + \dfrac{250}{10} + 15 = 56 \text{ kN/m}^2$

(iii) By stiffness $\quad P = 25 \times 2 \times 1.5 + 5 = 80 \text{ kN/m}^2$

The figure of 56 kN/m^2 is selected as arching effect produces a favourable result in such a thin section.

USA formulae

$$P = \frac{150 + 9000 \times 2 \times 3.24}{59} = 1138 \text{ lb/ft}^2$$

$$= 1138 \times \frac{47.88}{1000} = 55 \text{ kN/m}^2$$

Chapter 30 PRESTRESSED CONCRETE

INTRODUCTION

Normal structural concrete members rely on the reinforcing steel to take up tensile stress as load is applied, allowing the surrounding concrete to crack when its tensile limit is exceeded. Such an effect is particularly undesirable in water-retaining structures and those subject to alternate freezing and thawing. To overcome this problem the concrete can be precompressed such that tension is just obtained when the design load is applied.

Today precompressing (more commonly referred to as 'prestressing') has been extended to general applications such as bridge beams (fig. 30.1) and decks (fig. 30.2) foundations and piles (fig. 30.3) subject to impact loading, suspension bridge anchorages (fig. 30.4), rock bolting (fig. 30.5), portal frames (fig. 30.6) and for specialised work such as dam strengthening (figs. 30.7 and 30.8).

Figure 30.1 Simply supported beam

Figure 30.2(a) Continuous deck with tendon coupling

Figure 30.2(b) Bridge deck with tendon couplings

Figure 30.3(a) Basement anchors

Figure 30.3(b)
Foundation anchors

Figure 30.4 Cables
anchorage

Figure 30.5 Rock
bolting

Figure 30.6 Portal
frame connections

Figure 30.7 Dam
strengthening

Figure 30.8 Retaining
wall

253

DESIGN PRINCIPLE

The principle of prestressing requires the concrete to be initially forced into a state of compression which is then counteracted by the working load placed on the member. For example a beam subjected to bending would be stressed as shown in fig. 30.9, on applying a uniform compressive stress of equivalent magnitude over the full face of the beam, the bottom tensile stress would be nullified while the top compressive stress doubled.

Figure 30.9 Stresses under a symmetric prestressing force

Figure 30.10 Stresses under an eccentrically applied prestressing force

Furthermore, it can be seen in fig. 30.10 that by applying the prestressing force eccentrically a 'hogging' effect is produced, thus relieving some of the compression in the top fibres. In practice, the member will sometimes only be loaded with its own weight and therefore stability is necessary in both this condition and under live load.

The four cases of stress which must be kept within the critical strength limits of the materials used are as follows:

1. The bottom compressive stress in the concrete under self-weight.
2. The top tensile stress in the concrete under self-weight.
3. The bottom tensile stress in the prestressing tendon under self-weight and live load.
4. The top compressive stress in the concrete under self-weight and live load.

TENDON PROFILE

In the case of beams produced by the 'long line' process the tendon is cast horizontally in the lower section of the member. The bending moment,

Figure 30.11 Bending moment diagram with a uniformly distributed load

however, will be parabolic as shown in fig. 30.11. Consequently the eccentricity of the applied force may cause the critical stresses to occur as described in (2) and (4) above, near the end supports where there is little bending. This effect can be reduced by choosing a post-tensioning method whereby the eccentricity of the cable is altered to match the reduction in bending moment as shown in fig. 30.12 for a simply supported member. Thus at positions of zero bending moment the prestressing force can be arranged to produce uniformly distributed compressive stress across the face. To achieve this the cable is positioned in the member by means of ducting firmly attached to the shear reinforcement stirrups and cage. Joints in the ducting must be well taped to avoid penetration of the grout during subsequent filling of the void.

Figure 30.12 Cable in parabola to match eccentric prestressing force

METHODS OF PRECOMPRESSING (PRESTRESSING)

High tensile steel (HTS) bars, wires, or strands are stretched, and anchored at the ends of the concrete member. Alternatively, the concrete may be cast around the tendon to provide bond between the steel and concrete after hardening. The latter is usually the basis of pretensioning and the former post-tensioning.

MATERIALS

STEEL

Many materials have been tried for prestressing, but cold drawn high tensile alloy steel wire or cold worked bar have proved the most reliable, both being made from hot rolled rod. The tendon itself may be in the form of a single wire or bar, a strand made up of a bundle of wires spun together, or cable, which is a bunch of strands.

STRAND

A strand usually comprises seven wires (fig. 30.13), occasionally nineteen, six being hellically wound on to a straight single wire core. Typical performance characteristics of both strand and cable are given in table 30.1.

Figure 30.13 Strand configuration

NORMAL OR STANDARD   COMPRESSED (SUPER OR DRAWN)

BAR

High tensile steel bar is obtainable smooth or ribbed and supplied in lengths up to 18 m. Available diameters are usually in the range 20–50 mm. The corresponding permissible prestress loads at 70% of the characteristic failing load are 228–875 kN. Threads are *rolled* on at the ends of the bar for anchorage or coupling purposes. The ribbing, however, will satisfactorily serve the purpose in the case of the ribbed bar.

Table 30.1 Typical Design Forces
Strand, comprising seven wires

Strand diameter (mm)	Characteristic strength (kN)
12.5	165
15.0	225
18.0	370

Cable, comprising seven-wire strand

Strands/strand diameter (No.) (mm)	Permissible tension Force (i.e. 70% characteristic strength (kN) for standard strand)*
4(12.5)	450
4(15)	650
4(18)	1050
7(12.5)	800
7(15)	1150
7(18)	1800
12(12.5)	1400
12(15)	1950
12(18)	3200

* Higher tensile strength can be obtained with super and drawn strand.

PERMISSIBLE PRESTRESS LOAD IN THE TENDON

Unlike ordinary mild steel, cold drawn or worked high tensile steel does not have a clearly discernible yield point (fig. 30.14) and thus some other method of determining the permissible prestress load is needed. The British Standard uses the concept of a proof load (fig. 30.15), this being defined as the load which produces a particular permanent increase in the length of the wire – 0.1% is the commonly adopted elongation. The load is determined by drawing a line starting at the 0.1% elongation point and proceeding parallel with the elastic section of the graph until the deformed point of the curve is reached – this point is the 0.1% proof load, and in practical terms is approximately 80% of the characteristic failing load of the specimen. The applied prestress tension should normally fall within this value. It should be noted that wire or strand is nowadays generally supplied as 'relaxed', i.e. the curvature resulting from the drawing process is straightened by stress-relieving heat treatment.

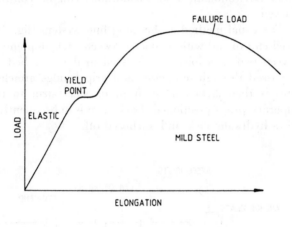

Figure 30.14 Mild steel tensile load/extension characteristics

Figure 30.15 High tensile steel load/extension characteristics

CONCRETE

The concrete for prestressed concrete applications should demonstrate low shrinkage, creep and elastic deformation, combined with high strength in compression. These characteristics can be obtained by using a low water/cement ratio together with selection of good quality and well graded aggregates. Precast concrete units commonly do not require congested cages of reinforcement and low workable mixes can be satisfactorily handled.

PRETENSIONED CONCRETE METHOD

High tensile steel wires are tensioned between two anchorages and then surrounded by concrete in a formwork mould. When the concrete has gained enough strength the ends of the wires are released from the anchorages. Immediately the wires seek to return to their untensioned length, but being restrained by the bond with the hardened concrete, compressive force is transferred in the member. Some elastic deformation takes place, and slowly over time both shrinkage and creep will act to reduce the tension in the wires and thus the level of compression in the concrete.

While pretensioning is suitable for making individual or specialised units the most common application is in mass production of standard items in a factory environment, where conditions can be controlled and high quality achieved.

The usual method is the 'long line' system (fig. 30.16). The moulds are placed end to end with a space between each stop end and the wire stretched between two steel joists firmly fixed to the floor bed. The ends of the wires are passed through templates and grip fittings attached. One of the anchor plates is then jacked away from its steel strut to take up the slack and temporary props positioned. Each wire is subsequently tensioned as required with a hydraulic jack and anchored off.

Figure 30.16 Long line method of casting

When the concrete has reached sufficient strength the exposed wires between the units are cut. The stress at these points will suddenly be reduced to zero, rapidly increasing to the prestress tension as the bond takes effect. As a result lateral swelling occurs to form an ideal cone shaped anchorage. Generally shearing reinforcement in the form of stirrups is required along this short transfer length of tendon.

POST-TENSIONED CONCRETE METHODS

The concrete may be cast either as the complete unit or alternatively in segments before the tendon is stressed. Usually the tendon is introduced after the concrete has hardened by casting in duct tubes arranged to give the appropriate profile (fig. 30.17). Occasionally, for example when using bar, the tendon is cast with the concrete either in a sheath wrapped in tarred paper

Figure 30.17 Duct profile

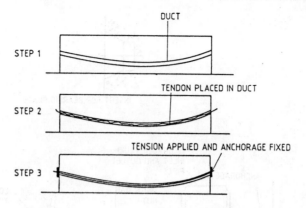

or with a bitumen coating to prevent bonding. When the concrete has reached the 28 day strength post-tensioning can generally proceed, using one of several proprietory systems.

With the exception of bar they are all very similar and differ mainly in the form of anchorage and arrangement of the tendon into wire, strand or cable.

BAR (fig. 30.18)

The two common systems are:

(i) Macalloy bar supplied by Reinforcement Steel Services PO Box 71 Hawke, Sheffield – smooth bar.
(ii) Dividag bar supplied by Dividag Stressed Concrete Ltd Westfield Road, Southam, Leamington Spa – ribbed bar.

The usual form of anchorage transfers the tension in the bar through a nut and washer to a solid plate, bearing against the concrete.

Figure 30.18
Threaded bar and nut system

TENSIONING PROCEDURE

The jack (fig. 30.19) comprises a hydraulic pump powered by a diesel motor. The anchor nut-tightening device is integral with the jack housing and consists of a hexagon pinion manually turned using a ratchet spanner.

Before commencing tensioning, the end plate washers and anchor nuts are positioned. The hexagon insert is placed over one of the nuts and the draw bar screwed on to the projecting bar thread. The jack is then slid on to this bar and the draw bar nut screwed against the ram. A small amount of load

Figure 30.19
Tensioning jack for the
bar system

INSERT FOR BEARING PLATE

is applied to take up slack and looseness and the initial dimension x measured. Further load is subsequently applied and the anchor nut tightened until full load is reached. The increased value is measured and compared with that expected for the actual load. Any difference outside the manufacturer's quoted tolerance should be referred back for advice.

On completion the jack is removed and the duct grouted as a precaution against corrosion of the tendon.

BARREL AND WEDGE GRIP SYSTEMS

The majority of manufacturers have adopted strand systems rather than bar. The anchoring method generally consists of barrels and conical grooved wedge grips (fig. 30.20a) butting against a strong cast iron distribution plate cast in the concrete. Alternatively, the conical wedges may directly engage the anchor block as shown in fig. 30.20(b).

The wedge, tapered on the outside, is usually in two or three segments held together with a wire band collar and is grooved on the inside to aid grip with the tendon. Where several strands are involved, each is passed through the anchor block to produce a regular symmetrical pattern, the strands being gathered together to form a cable all housed in a single duct.

To achieve post-tensioning, the strands are either stressed individually in the order specified by the manufacturer, or more commonly all together as a cable. In the former case the jack is relatively light and man-handleable, for the latter because of its weight. Expensive lifting assistance is required, but the stressing time ought of course be shorter. The typical stressing procedure shown in Fig. 30.21 requires a single strand to be passed through the jack, and either a temporary barrel and wedge fixed behind the ram, or alternatively the tendon is gripped inside the jack. The slack in the grips and

Figure 30.20 Barrel and wedge systems of anchorage

Figure 30.21 Tensioning jack for single-strand barrel and wedge system

261

Figure 30.22
Tensioning procedure

Figure 30.23
Tensioning jack for
multi-strand barrel and
wedge system

tendon is then taken up (extension e_1 and e_2 in fig. 30.22) and the required load applied. The extension of the strand e_3 is measured and compared with expected value. Any significant difference should be referred back to the manufacturers for advice. If everything is satisfactory, a ram inside the jack is activated to push the wedge into its barrel. The stressing force is then released causing the wedges to tighten, and increase the grip on the strand. A small loss of tension will occur during this phase (e_4). The temporary wedge and barrel can then be removed and the jack transferred to the next strand. A similar procedure can be adopted for stressing the strands simultaneously (fig. 20.23). Single-strand jacks are available in the range 50–300 kN and 50–500 mm strokes, cable-stressing jacks up to 10 000 kN and bar jacks 700–1600 kN.

Many systems are available from different manufacturers all very similar. A few of the more popular ones are listed below.

1. CCL Systems Ltd, (fig. 30.24)
   Ewell Road, Surbiton, Surrey.

   Monostrand and multiforce systems.

   Strands stressed singly or simultaneously.

Figure 30.24 CCL system

2. Losingers Systems Ltd, Lupton Road, Tame, Oxon. — VSL. — Strands stressed simultaneously.

3. PSC Fregssinet Ltd (fig. 30.25), The Ridgeway, Iver, Bucks. — Monostrand and monogroup systems. — Strands stressed singly or simultaneously.

4. Pilcon Engin, Ltd (fig. 30.26), Alencon Link, Basingstoke, Hants. — SCD monogrip and multigrip systems. — Strands stressed singly or simultaneously.

5. Stronghold International Ltd, Oxford Road, Stone, Aylesbury, Bucks. — Stronghold system. — Wire or strands stressed simultaneously.

Figure 30.25 PSC system

Figure 30.26 SCD system

THREAD AND NUT WITH WIRE SYSTEM

Simonbuild Ltd, Cheadle Heath, Stockport, market the BBRV system (fig. 30.27) Individual wires are passed through the anchor head and button ended. A choice of two types of anchorage is available – the anchor head is either threaded internally to receive a temporary pulling bar (a), or threaded externally to receive a pulling sleeve (b), the latter being more appropriate for larger anchorages. The tensioning method is similar to that of the bar system and thus the tendon length needs to be accurately determined to allow for the extension. In the final condition the tension is transferred from the anchor head on to a distribution plate.

Figure 30.27 BBRV system

COUPLINGS

Occasionally tendons require coupling as demonstrated in fig. 30.28. Where a bar is used the method is quite simple requiring only a threaded sleeve (fig. 30.28), but for strand, usually the tendons are arranged around integral anchorage and counter anchorage heads (figs. 30.29 and 30.30).

The various manufacturers generally have purpose made systems for such joints.

Figure 30.28 Bar coupling

Figure 30.29 Wedges and strands coupling

Figure 30.30 BBRV
wires coupling

LOSSES AFTER STRESSING

A loss in tension in the wires takes place as a result of:

			Loss %
(i) Shrinkage of the concrete		up to	5
(ii) Creep of the concrete		up to	10
(iii) Elastic strain in the concrete		up to	5
(iv) Creep or relaxation of the steel		up to	5
(v) Losses during the tensioning and anchoring operations		Total	20–30%

SHRINKAGE

As concrete hardens, shrinkage takes place, partly caused by the chemical reactions and partly by drying out. Practice has demonstrated that low water/cement ratios produce less shrinkage, for example

Water/cement ratio	Shrinkage (mm/mm of length)	
0.4	200×10^{-6}	approximately 5% of initial prestress
0.7	800×10^{-6}	

About half of the shrinkage takes place in the first 3 weeks and is virtually complete after 6 months.

CREEP

When concrete is held under a sustained compressive stress, permanent deformation occurs causing a reduction in length. The amount of reduction is proportional to the stress in the section and increases with time. Up to 10% loss of stres may be caused by creep.

ELASTIC STRAIN

As load is applied to the concrete a reduction in length of the member occurs, the amount being related to the modulus of elasticity of the steel and concrete

Figure 30.31 Creep and elastic deformation

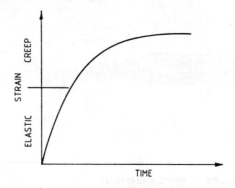

and is directly proportional to the stress applied. Elastic deformation takes place before creep (fig. 30.31) with the latter eventually reaching a maximum value. In pretensioned concrete, stress is immediately transferred to the concrete and may eventually result in up to 4–5% loss of prestress. In post-tensioned work, where the stress is applied in stages, strand by strand, progressive shortening occurs, causing some loss of stress in the previously tensioned tendons, about 2% loss of prestress is normally assumed. Where post-tensioning is applied to the strands simultaneously at one of the anchorages, then the effect of elastic deformation is nullified, since the force in the wire is under direct control at the jacking position. Thus no allowance is needed for prestress loss.

RELAXATION OF THE STEEL

Tests have shown that steel under a tensile load looses some of its stress over a period of time. The rate of loss decreasing with time such that over a period of about 1000 hours the limit is virtually reached. The amount of creep is greater for wires in the 'as drawn' condition (i.e. delivered in coils with a natural curvature relating to the capstan size of the wire-drawing equipment) compared with heat treated straightened wire. Up to 5% should be allowed for steel relaxation, when the actual stress in the tendon is about 70% of the breaking load (normal recommendation). Slightly less relaxation is exhibited by bar.

LOSS BEFORE AND DURING TENSIONING

In post-tensioned work, friction between the tendon and the duct prevents the full tension at the jacking point reaching positions further along the duct especially where curved profiles are involved, the degree of seriousness depending upon the duct material. Some loss of prestress also occurs on transferring the load to the anchorage.

LOSS OF STRESS CAUSED BY JOINTING

In post-tensioned work, the final member may comprise several smaller precast units butted together (fig. 30.32). To avoid movement problems after

Figure 30.32
Assembled precast post-
tensioned units

stressing, the units should be cast back-to-back, numbered and erected in an identical sequence.

Alternatively the member can be temporarily supported in the final working position and in the case of a wide joint, filled with *in situ* concrete (fig. 30.33). For a narrow gap high strength quick setting resin may be suitable.

Figure 30.33 Joint
movement with post-
tensioned units

BIBLIOGRAPHY: SECTION 2

1. *Aggregate: The way Ahead*, (1976) Report on the advisory committee on aggregates, HMSO.
2. Day, D. A. (1973) *Construction Equipment Guide*. Wiley.
3. Havers, J. A. and Stubbs, F. W. (1971). *Handbook of Heavy Construction*. McGraw-Hill.
4. Jackson, N. (1980) *Civil Engineering Materials* (2nd edn). Macmillan.
5. Peurifoy, R. L. and Ledbetter, W. B. (1985) *Construction Planning Equipment and Methods* (2nd edn). McGraw-Hill.
6. Sinclair, J. (1969) *Quarrying, Opencast and Alluvial Mining*. Elsevier.
7. *Pumping Concrete* (1985) Concrete Society, Digest No. 1.
8. Bamfort, P. B. (1985) *Mass Concrete*. The Concrete Society.
9. Deacon, R. C. (1985) *Watertight Concrete Construction*. Cement and Concrete Association.
10. Allen, R. T. L. (1985) *Concrete in Maritime Works*. Cement and Concrete Association.
11. Deacon, R. C. (1985) *Concrete Ground floors: their Design, Construction and Finish*. Cement and Concrete Association.
12. Allen, A. H. (1985) *The Repair of Concrete Structures*. Cement and Concrete Association.
13. Clear C. A. and Harrison, T. A. (1985) *Concrete Pressure on Formwork*. CIRIA Report 108.
14. Neville, A. M. (1987) *Concrete Technology*. Longman.
15. Final Report of the Advisory Committee on Falsework (Bragg Report) (1975).
16. *Formwork – a Guide to Good Practice*. (1986) Concrete Society.
17. Harrison, T. A. (1987) *Formwork Striking Times – Methods of Assessment*. CIRIA Report 73.
18. Abeles, P. W. and Bardhan-Roy, B. K. (1981) *Prestressed Concrete Designer's Handbook* (3rd edn). Cement and Concrete Association.
19. Allen, P. W. (1978) *An introduction to Prestressed Concrete*. Cement and Concrete Association.
20. Bate, S. C. C. *et al* (1979) *Handbook on the Unified Code for Structural Concrete* (CP 110:1972). Cement and Concrete Association.
21. Bate, S. C. C. and Bennett, E. W. (1976) *Design of Prestressed Concrete*. Surrey University Press.
22. Andrew, A. W. and Turner, F. H. (1985) *Post-tensioning Systems for Concrete in the UK 1940–1985*. CIRIA Report 106.

SECTION 3   ROAD PAVEMENT CONSTRUCTION AND EQUIPMENT

Both bituminous and concrete laying equipment are described, including modern paving trains. Batching, mixing and recycling of macadam and asphalt are also included. Some of the processes and equipment used in soil stabilisation are considered.

CHAPTER 31 FLEXIBLE PAVEMENT CONSTRUCTION

INTRODUCTION TO ROAD PAVEMENTS

A highway or road is composed of layers of selected materials which act to transfer and distribute vehicle loads to the subgrade. The objective being to limit the stresses within the capacity of the subgrade. This can be achieved with either a flexible or rigid pavement (fig. 31.1).

Figure 31.1 Flexible and rigid pavement layering

The flexible method relies on distributing the stresses by means of the pavement thickness, whereas a rigid pavement comprises a concrete slab utilising bridging action to span weak areas of the subgrade. As a consequence flexible pavements are commonly much deeper.

FLEXIBLE PAVEMENT DESIGN

The soil foundation formed after either completion of excavation works for a cutting or the formation of an embankment is called the 'subgrade'. Usually the surface of the subgrade is referred to as the 'formation level'. The flexible pavement is then built up in layers as follows:

SURFACE COURSE

The surface course is usually constructed to produce a comfortable ride for traffic, and should ideally be hard wearing, resistant to weather attack, and display good properties with respect to skidding. In addition adequate water-proofing to the lower layers is essential and can be obtained by careful selection of surface materials.

On highly used roads it is common practice to separate the surface course into two layers – wearing and base layers. This method of construction facilitates maintenance, repairs and, in particular, replacement of the wearing layer. However, for minor roads the two are usually combined. The surface courses are generally made from tar or bitumen-bound aggregate.

ROADBASE

The roadbase acts to distribute the wheel loads through to the subgrade and therefore performs a most important function in the design of the pavement. High quality materials are required, such as hardcore, graded granular materials laid either wet or dry, cement or bituminous stabilised soil or gravel, or lean mix concrete. Depending upon the material, the method of laying can either be carried out using a paving machine or by simple spreading with bulldozers or graders combined with adequate compacting.

SUB-BASE

This is an extension of the roadbase and will be required depending upon the quality of the subgrade and loading on the pavement. Compacted granular materials to aid drainage are common and may be of lesser quality than those used in the roadbase. Nevertheless the sub-base must be stronger than the subgrade.

RIGID PAVEMENT DESIGN

SURFACE SLAB

This forms a structural plate and is formed from high quality strong concrete, sometimes reinforced with steel bars or mesh. Common practice often allows for expansion and contraction by incorporating joints at intervals passing through the slab. Laying is generally carried out with a paving machine.

SUB-BASE

When the subgrade cannot provide uniform support to the slab because of the effects of frost action, poor drainage, swell and shrinkage, bad practice during earthmoving operations, construction traffic, etc., a sub-base is usually interposed between these two surfaces.

The sub-base may be formed of granular materials or a weak-mix concrete with the top surface sealed with a thin bituminised layer to prevent water absorption from the concrete layer. Laying can be carried out from a paver or simply by spreading, grading and rolling with earthmoving equipment.

SUBGRADE

The subgrade forms the natural foundation to the pavement and is the result of either excavation works or the forming of an embankment. Thus it is important that adequate compaction takes place before constructing the other layers.

Maximum compaction is achieved at some optimum water content for the particular soil (fig. 31.2). This value may be determined in the laboratory, with the subsequent standard of compaction to be achieved on site specified at some proportion of this value (Proctor test).

Figure 31.2
Compaction curve

The design thickness of the pavement may be determined, by several methods, for example California Bearing Ratio (CBR) of the soil. The laboratory test uses a disturbed sample which is re-formed and compacted in a standard mould to a state anticipated at the construction site. A penetration load test is then performed and the CBR value obtained. A similar procedure could be carried out *in situ* on the actual subgrade.

BITUMINOUS LAYING METHODS

Macadams and sometimes bituminous materials can be spread by hand and raked level before trial rolling. This technique, however, is only suitable for small scale works such as patching, footpaths, domestic work, etc. For road surfacing a paving machine is commonly selected.

THE PAVING MACHINE

Paving machines were introduced during the 1930s and are capable of laying a wide range of material consistencies and thicknesses. The width of the machine may be altered to suit the road design and its production potential is considerably greater than manual methods.

The paving machine comprises two parts, the tractor unit and the screed. The tractor, mounted on either tracks or wheels, supports the engine, transmission, controls, receiving hopper, feeders and spreader augers. The screed

Figure 31.3
Bituminous paving
machine

is pivoted at the rear of the tractor system and is simply towed on long
levelling arms. The screed unit carries the compacting tamper, thickness
adjuster, heaters and screed plate.

Laying Action

Paving material is dumped into the receiving hopper from trucks positioned
in front of the moving machine. It is subsequently conveyed to the spreading
augers, deposited, tamped (variable 0–25 cycles/sec) and then passed under
the screed plate.

The screed is hung from pivoted levelling arms. As the tractor pulls the
whole unit forward, the plate lines up parallel with the direction of pull (fig.
31.4). The depth of the material may be changed by adjusting the angle of
the plate.

However, because the levelling arms are quite long, the reaction will not
be immediate, and approximately 3–4 m of travel is required before equilibrium condition is achieved. This has the advantage of reducing the effect of
irregularities which the tracks or wheels may meet, and therefore the finished
surface will be much smoother than the travelling surface.

Figure 31.4 Laying
action

Automatic screed control

The unpaved surface may have undulations and irregularities which could be undesirable if reproduced on the riding surface. Modern machines, however, are equipped to adjust the pivot height of the screed with respect to a guide line, thereby allowing the shape of the surface to be controlled.

There are three main types of automatic control system:

Stringline

A stringline accurately set out between support pegs acts as a guide for a sensor mounted on a pivot arm of the paving machine. Usually the stringline is only required at one side of the machine as a built-in crossfall control mechanism operated automatically or manually sets the height of the opposite pivot arm. Stringlines are not particularly suitable for producing a flat surface because of sag and the latest machines would alternatively be equipped with laser sensing devices to do a similar job.

Figure 31.5 Stringline control

Shoe

A kerb line or previously laid surface acts as the reference line for a 'shoe', which in turn activates the sensors on the pivot arm.

Figure 31.6 Averaging beam control

Figure 31.7 Matching shoe control

Averaging beam

The averaging beam comprises two pairs of feet, each pivoted to a stiff beam. The beam is independently pulled along, to match the travel of the paver, and a sensor on the paver's levelling arm is activated on touching the reference wire attached to the top of the beam. Thus the irregularities are bridged resulting in improved smoothness of the paved surface.

Wheels or Tracks

On undulating surfaces, tracks ride on the crests of the irregularities, and provide similar advantages to the averaging beam, where as wheels tend to reproduce the shape of the riding surface. However, with a bump such as a manhole cover the wheeled version produces only a small isolated deflection of the screed, while with tracks, the effect is reproduced along the full length of the tracks as the machine passes the irregularity. Also in the case of manholes it is sensible practice to lay material over the cover plate which can then be lifted flush later.

PAVING PRACTICE

1. The base surface should be clean, firm and free from excessive irregularities.
2. A tack coat of bituminous emulsion (0.5 kg/m^2) is required to give adequate adhesion to a non-bituminous based surface, or to prevent water seepage if the base course is of porous material.
3. Work uphill to give better feed control for the lorries.
4. When using a hot mix some heating of the screed prevents material sticking to the plates of the paving machine.
5. Run the paver at a travelling speed to match the rate of delivery of material, frequent stopping and starting may produce surface irregularities.
6. A thin wearing layer should not be laid on a wet surface if good adhesion is to be achieved.
7. Never allow the roller to remain stationary on the compacted surface, otherwise dents, wheel ruts, etc. may occur.

LAYING WIDTH

Pavers are manufactured with an approximate 3 m standard width screed, but extension augers and screeds are available to produce a carpet up to 12 m

Figure 31.8 Laying width adjustment

wide. Common practice is to lay a third or half the pavement width, then turn the machine around and lay the next strip. This keeps the cost of the operation down (fitting extensions is expensive) and allows traffic to keep flowing without diversions. The main disadvantages are the risk of damage to the edge of the newly laid carpet by passing traffic and faulty jointing of the strips.

PAVING DATA

Paving widths	3–12 m
Paving thickness	up to 300 mm
Paving speed	1–30 m/min
Travelling speed	25 km/h
Output up to	500 t/h
Bearing pressure under tracks	0.05–0.15 N/mm^2
Bearing pressure under tyres	0.05–1 N/mm^2
Transport of hot mixes	not longer than $\frac{1}{2}$ h

COMPACTION

Traditionally, adequate compaction has been assumed by specifying the method of compaction. This is described as an 8–10 mg smooth wheeled static roller, with a roll width of not less than 450 mm or a multi-wheeled pneumatic tyred roller of the same weight, the surface layer always being rolled by a smooth wheeled type.

Rolling generally proceeds longitudinally, starting from the side working toward the centre of the strip, overlapping the preceding run by approxi-

Table 31.3 Compaction methods

Roller type	Action	Advantages	Disadvantages
Static weight	Static weight	Produces a smoother, flatter surface with increasing stiffness of the material. Gives a smooth even surface	Effective only for shallow thickness
Vibration rollers	Static weight and dynamic action	Effective compaction for thick layers Good interlocking of aggregate	Difficult to determine the most suitable frequency and speed of travel
Rubber tyred rollers	Static weight and kneading action	Reasonably good interlocking locking of aggregate Effective compaction for thick layers Good surface seal Ability to vary contact pressure Uniform compaction	Expensive to purchase Experienced driver needed

mately half the roller wheel width. Rolling continues until all the roller marks have disappeared. The roller wheels must be kept moist to prevent adhesion. In more modern practice, however, the resultant density is specified to a percentage value of that obtained during laboratory design tests. This has caused investigations into more efficient compacting methods, especially using vibration and has allowed the specification of thicker construction layers. The principal rolling methods are summarised in table 31.1.

EFFECT OF TEMPERATURE

Ideally the temperature of the material at the time of compacting should be approximately 50–60 °C above the softening point of bitumen (ring-and-ball test). For most of the common types of bitumen and tar, softening occurs at about 25 °C, thus compaction should have been completed before the materials temperature falls to 80 °C or so. Since the mixing temperature for asphalt and macadam is between 100–150 °C, during cold weather little time is available to achieve good compaction results. Figure 31.9 illustrates the approximate time needed to compact various thicknesses when the outside temperature is near freezing. It can be seen that for thin layers cooling is very rapid. Hot boxes can be used to maintain the temperature at a high level for several hours, but evaporation of the oils can result in segregation of the aggregates.

Figure 31.9 Effect of temperature on compaction time

NUMBER OF COMPACTORS

The required number of passes to achieve compaction is usually determined in the laboratory and the required density subsequently specified for the construction phase. However, the conditions for achieving acceptable compaction results depend greatly upon the temperature of the material. For example, rolling of hot, soft material, especially asphalt, produces tracks, while cool harder material is difficult to compact. Thus the size and number of compactors must be carefully balanced with the capacity of the paving machine, and the time available before cooling is complete.

SURFACE DRESSING

As a road gets used the surface may suffer degradation due to a combination of traffic and weather conditions. Ultimately the wearing layer may have to

Figure 31.10 Surface
dressing cross section

CHIPPINGS
BINDER FILM
EXISTING PAVEMENT

CHIPPINGS SPREAD ON SURFACE

INITIAL EMBEDMENT INTO BINDER FILM

FINAL EMBEDMENT INTO EXISTING PAVEMENT

be completely removed for example on major highways, or alternatively be given a new surface dressing as is commonly adopted on secondary roads and housing estate streets.

The surface dressing process consists of the application of a film of bituminous or tar binder on to the existing road surface followed by a layer of stone chippings to be subsequently rolled in. The process gives:

(i) Improved non-skid wearing surface.
(ii) Waterproofing of existing surface.
(iii) Arrests surface disintegration.

CHIPPINGS

In order that adequate adhesion to the binder can be obtained, the chippings should be strong, durable, dry and free from dust, otherwise complete failure of the surface may occur on heavily used roads. An adhesion agent can be added to reduce this risk.

APPLICATION OF SURFACE DRESSING

The surface must be clean and free from: loose fragments, stone, dust, mud, etc. Also soft spots and depressions should be made good before the surface dressing is applied.

The bitumen or tar binder is sprayed uniformly across the old surface at a rate dependent upon the size and shape of the chippings, the hardness of the surface and the expected traffic density. The spraying equipment comprises a mobile tank fitted with pressure and rate of flow controls, together with a temperature monitor.

Common practice is to lay surface dressing equal to one traffic lane width, while additional lanes may be overlapped or butt jointed. The chippings are usually spread mechanically (fig. 31.11) at a rate dependent upon the size,

Figure 31.11 Surface dressing spreader

shape and specific gravity of the stones. A generous quantity is spread completely to cover the binder film. The final operation involves rolling the chippings into the old surface, preferably with a rubber-tyred or rubber-rimmed roller to avoid crushing.

REPAVING

The removal of the wearing surface can be achieved with either hot or cold planing techniques, and the material reprocessed in a heater (drum mixer) and used again if necessary.

HOT PLANER

The equipment comprises a self-propelled or pulled wheel-mounted burner and separate cutting drum. The layer to be removed is heated by means of propane-fired infra-red heaters, the softened material subsequently being scraped away with the planing drum to leave a relatively smooth, even surface for repairing. The loosened material is usually limited to about 40 mm working depth. The machine is rather cumbersome weighing roughly 30 tonnes.

Figure 31.12 Hot planing machine

COLD PLANING

Cold planing is becoming increasingly popular, because of its ability to cut to greater depths (typically up to 300 mm) than the hot technique and produces no fumes.

The equipment simply comprises a wheeled or tracked cutting drum and elevator to transport the material to trucks for disposal. The planing action

279

Figure 31.13 Cold
planing machine

is achieved with tungsten carbide tipped teeth. Drums capable of spanning widths up to approximately 4 m are available. Like paving machines, automatic slope and grade controls can be included to remove the pavement to ± 3 mm or so. The power unit is commonly in the range 250–600 kW with a production capability of 300–800 t/day, working at up to 50 m/min for shallow cuts of 25 mm. However, speeds as low as 5 m/min are not unusual for deep cuts. Such machines lie in the weight range 20–50 tonnes.

RECYCLING OF ASPHALT SURFACE MATERIAL

Recycling of existing surface material is becoming increasingly common as the cost of bitumen products rise. The technique, however, dates back to 1940s and takes place as either a hot or cold process, using material obtained from the planing process. Recycled pavement has been laid as base courses, and top or wearing courses for secondary roads.

HOT RECYCLING

The more commonly used equipment is a drum mixer, similar to fig. 32.8. A 'mid-entry' method is favoured whereby the reclaimed asphalt pavement is introduced downstream from the burner and mixed with the heated new aggregate which has passed through the flame. After adding sufficient binder and mixing, the material is transferred to a storage bin.

A reclaimed asphalt/fresh aggregate ratio of 70/30 is achievable, but 50/50 is the usual practical limit to avoid air pollution problems – mixing temperatures exceeding 140 °C (280 °F) produces unpleasant quantities of blue smoke. The appropriate mixing temperature depends largely on the moisture content of the reclaimed material. The finished material is usually laid with conventional paving plant.

COLD RECYCLING

Cold recycling is generally an in-place method using material obtained from cold planing to which an admixture or bitumen rejuventator (e.g. foamed bitumen) is added. To achieve successful results at least 95% of the planed material should pass through a 50 mm screen.

A common method of working involves spraying the pulverised surface with the additive followed by mixing during a second pass of the machine. However, some modern equipment allows the additive to be introduced during either the grinding or mixing phases (fig. 34.6). Subsequently reshaping and compacting can be performed using conventional rollers. The technique is more appropriate to secondary roads with the recycled surface subsequently given a sealing of chippings or new wearing surface. A more conventional method of recycling requires distribution of the blended material as a windrow behind the planer to be picked up by a paving machine and re-laid in the normal manner or spread by bulldozers or graders. However, the best results are usually achieved by a travelling pugmill mixer/paver machine, fed from strategically positioned stockpiles of material crushed to the desired size with pulverising equipment (hammermill). Any additive is blended in the pugmill, the contents thoroughly mixed and relaid to grade and slope. The method is suitable for base courses up to 350 mm deep.

Chapter 32 BITUMINOUS-BASED MATERIALS FOR FLEXIBLE PAVEMENTS

The surface layers and often the road base of a flexible pavement are made from a combination of a binder material and aggregates. There are several mixtures in common use as outlined below.

HOT ROLLED ASPHALT

This consists of fine aggregate and a bitumen binder to which various proportions of coarse aggregate may subsequently be added to suit the particular circumstances. For example 30% coarse aggregate (i.e. that returned on a 2.36 mm sieve) is typically used for the wearing course, but higher proportions are commonly selected for base courses and road bases.

Coated chippings are generally rolled into asphalt wearing surfaces to improve skid resistance.

DENSE TAR MACADAM

Tar is used as the binder and laid as a hot material. Like hot rolled asphalt the proportion of coarse aggregate may be varied. Rolling produces a dense, impervious surface.

COATED MACADAMS

These contain little fine aggregate and either tar or bitumen may be used as the binder or coating agent. As a result the material has an open texture and depends on an interlocking of the aggregate particles for strength and stability. This method is suitable for all-weather roads lightly trafficked, garage forecourts, footpaths, etc.

Coated macadams may be laid hot, using a high-viscosity binder, or cold with a low-viscosity binder, such as 'cutback bitumen'. The latter method allows the material to be stockpiled and used at intervals as required without detrimental effect.

PRODUCTION OF ASPHALTS AND MACADAMS

The plant required to manufacture asphalts (fig. 32.1) and macadams (fig. 32.2) basically provide the following operations:

(i) Drying the aggregate and removing dust.
(ii) Screening the aggregate into the required sizes and sorting into storage bins.
(iii) Heating the aggregate to the appropriate temperature.
(iv) Heating the binder to the appropriate temperature.
(v) Providing fine filler material.
(vi) Weighing, batching and mixing.

Figure 32.1 Asphalt plant

Figure 32.2 Macadam plant

Asphalt contains a high proportion of fines, i.e. sand, which tend to segregate from the coarse aggregate during movement after drying, while macadam being largely a coarse material does not suffer this effect. Thus for the asphalt-making process the screening, sorting and weighing is usually the last operation before mixing.

DRYING PROCESS

In order to obtain an efficient coating of the aggregate with the binder, the surface moisture content must be reduced by drying to less than 1% by weight. The aggregate is delivered by belt conveyor or feed hopper from the storage bins to the drier. The most common type consists of a rotating drum, fitted internally with lifter blades to cause a cascading motion as the aggregate is propelled forward. It is usual to have the drum on a slight downward slope. Heating is commonly provided by an oil burner located at the discharge end of the drum so directing the hot gases towards the oncoming aggregate. The high percentage of fines in an asphalt mix necessitates the drying unit to be much larger than that for macadam. Furthermore in the case of macadam, heating of the aggregate to the mixing temperature is carried out as a separate process and therefore the temperature in the drying process may be relatively low (a little above 100 °C).

HEATING THE AGGREGATE

For asphalts, heating and drying are performed in a single operation, after which the aggregate is screened into separate sizes and temporarily stored in small hot hoppers before being combined together and mixed with the binder. This is necessary to avoid the fines separating out during movement. Macadam, however, is stored in bins immediately after drying and then proportioned into the batch size and transported to a separate rotary heater drum for mixing.

HEATING THE BINDER

The bitumen or tar binder is heated in an insulated tank either with steam or hot oil circulating through pipes located inside the tank.

Drying and Heating Control Data

(i) Hot laid asphalt – drying/heating temperature:

Binder	*Aggregate*
135 °C (275 °F)	150 °C (300 °F) ± 10 °C

(ii) Cold laid asphalt – drying/heating temperature:

	Binder	*Aggregate*
Straight bitumen	125–160 °C	80–120 °C
Cutback bitumen	90–125 °C	50– 75 °C

(iii) Macadam (drying temperature 100 °C+); heating temperature:

	Binder	*Aggregate*
Straight bitumen	95–150 °C	65–150 °C
Cutback bitumen	65–130 °C	max. 70 °C

Capacities of Primary Driers

Production	50–500 t/h.
Oil consumption	5–15 kg/t (depending upon moisture content).

DUST EXTRACTION

Excessive dust may reduce the efficiency of the plant and in any case would cause difficult working conditions when the aggregate is moved from process to process. The dust is extracted during the drying phase using fans to direct it to a collector, and then removed by spraying with a stream of water. The resulting sludge is deposited at the base of the tank and subsequently removed and dumped. Approximately 95% of the dust can be removed in this way.

The cyclone collector (fig. 32.3) is a more modern process and gives a dry product which can be fed back into the system as filler material. The dust-laden gas enters tangentially at the top of a conically shaped drum, causing the particles to be thrown centrifugally on to the walls. A clean air vortex is consequently formed up the centre of the cone while the dust falls to the base

Figure 32.3 Cyclone collector for fines

Figure 32.4 Cyclone collectors in series

for collection. Approximately 60–90% of the dust may be removed, depending upon the particle size. Where a single unit is insufficient the cyclone collectors may be arranged in series (fig. 32.4). Furthermore a wet collector can be added as a secondary unit if necessary.

PROPORTIONING AND MIXING

The apparatus for screening and storing sand and aggregate is described in the section on aggregate production (see page 213). For macadam screening, storing and proportioning are carried out immediately after drying but before heating, while for asphalt this operation takes place after the combined drying/heating operation. Both weight and volumetric methods of proportioning are used, although for accuracy weigh batching is favoured. First the aggregate is drawn from the storage bins and transferred into a weigh hopper in the required proportions until enough material to produce a particular batch size has been weighed out. After weighing it is discharged into the mixer and the binder added. Proportioning of the binder may be weighed out similar to the aggregate but in a separate trough. Pumping the hot binder and monitoring through a flow meter is an alternative method, however this technique is unsuitable for binders containing abrasive matter such as lake asphalt.

Generally the last process in the sequence involves introduction of the filler, but unlike the aggregate and binder, is stored in silos and pumped or conveyed to a storage bin and subsequently weighed in a separate hopper before transfer to the mixer.

MIXING

A batch mixer consists of a steel drum, fitted with twin counter rotation (approximately 40 rev/min) parallel shafts supporting a series of paddles (fig. 32.5). The material is discharged through the base.

Continuous mixers are longer and the material enters at one end and discharges at the other, thus the paddles are designed to keep the mix rolling forward. It is customary to provide heating equipment to the casing. In batch mixing approximately 20–40 s mixing time per batch is usual.

Figure 32.5 Batch mixer

MODERN DEVELOPMENTS IN BITUMEN-COATED MATERIAL

PRODUCTION

The objective of the several new methods is to increase the rate of production and to reduce the dust problems associated with the older techniques. Figure 32.6 illustrates the use of the drying drum after the mixing process. The aggregates are screened and batched in the normal way and fed into the mixer, where the filler and hot bitumen are added. The mix is then transferred into the drying drum now called an 'activator', and the water in the aggregate is evaporated. An additive is generally included during the mixing phase to enhance the separation of the water and bitumen.

Figure 32.7 shows an alternative arrangement where the mixer unit is completely eliminated, mixing takes place in the drying drum together with the filler and bitumen. This method does not remove all the water, leaving a final moisture content of 1–3%. A binder additive is usually included.

These new methods are well suited to handle recycled asphalts and macadams (fig. 32.8).

Figure 32.6 Drying drum after the mixing process

Figure 32.7 Production plant with combined mixer/drying drum

Figure 32.8
Mixer/drying drum

Production Data: large units (Asphalt)

Output (t/h)	25	50–100	100–120	120–150	150–200	200–250	250–300
Drier size (m)	1 × 6	1 @ 8 × 2	2 @ 8 × 2	2 @ 8 × 2	4 @ 10 × 2	8 @ 10 × 2	10 @ 10 × 3
Mixer capacity (kg)	250	600–1000	1300	1500	2000	2500	3000
Filler silos (t)		1 × 25		2 × 25			5 × 60
Floor area (m²)	1000	1500	2000	3000	2500	2500	3000

Note: smaller plant down to 5 tonnes per hour is available. For macadam production a secondary but smaller heater is necessary.

Chapter 33   CONCRETE PAVEMENT CONSTRUCTION

Mechanised methods for laying concrete pavements were introduced during the 1930s but the popularity of bitumen products has largely overshadowed concrete. The techniques for forming concrete roads with mechanised equipment may be classified as:

1. Paving trains.
2. Slip-form pavers.

The need to give adequate compaction, careful surface finishing, and to cope with jointing and grooving has demanded more complicated pavers than those used for asphalt and macadam.

TRANSVERSE JOINTS

Joints in the pavement are required to facilitate expansion and contraction caused by temperature and moisture changes.

EXPANSION JOINTS

These are usually incorporated at about 100 m centres, while those for contraction are required at about 5–10 m centres. The type of expansion joint (fig. 33.1) may be the same for both reinforced and unreinforced slabs and comprises dowel bars cast into the concrete, with one half coated in bitumen and encased in a loose sheath to allow movement. Any reinforcement in the deck must be discontinued at the joint. During construction the bars are supported on fibrous boarding secured to the sub-base – the material being

Figure 33.1 Expansion joint

flexible to accommodate the expansion movement. The top of the groove is filled with joint sealent.

Recent developments in laying pavements during the summer months, however, have indicated that expansion joints can be eliminated as there is usually sufficient shrinkage at the contraction joints to allow for any subsequent expansion.

CONTRACTION JOINTS

These are usually induced by forming a groove in the top of the slabs with temporary filler bar (fig. 33.2). The resulting plane of weakness subsequently cracks thereby providing freedom for contraction. When the slab is unreinforced, dowel bars must be inserted to prevent vertical movement and like expansion joints, one half of the dowel must be free to move and is therefore coated with bitumen (fig. 33.3). When the concrete has reached sufficient strength the filler bar is removed and replaced with a sealing compound.

Figure 33.2 Contraction joint for unreinforced slabs

Figure 33.3 Contraction joint reinforced slabs

LONGITUDINAL JOINTS

Dowelled longitudinal joints are provided when the slab is concreted in more than one strip (fig. 33.5) or more than about 5 m wide. They serve to prevent the slabs drifting apart and/or differential vertical movement. When the road is laid as a single carpet and is wide enough to require joints, these may be formed in a similar manner to the previously described dowelled contraction joints but with the bars assembled in a continuous cage and placed in position ready for the concreting operations. Otherwise they generally need to be cranked through 90° tied to the side forms, and later straightened before concreting of the abutting slab.

REINFORCED PAVEMENTS

Concrete slabs reinforced with mesh are laid in two-course construction. The leading paver first places the bottom layer of concrete, followed by a work gang manually placing the sheets of mesh or (alternatively by machine). Finally the top course of concrete is placed by a second paving machine.

FIXED FORM PAVING TRAIN

Paving train equipment rides on rail track mounted on steel side formwork positioned along the pavement edges (Fig. 33.4). The forms are usually fixed in position a day or so in advance and not removed for several hours after concrete placing. Thus sufficient length must be in position to meet the planned rate of advance of the paving training.

The main units and sequence of operations in forming two course construction are shown in fig. 33.5.

Figure 33.4 Edge formwork for use with paving train system

Figure 33.5 Paving train equipment

SEQUENCE OF OPERATIONS

1. Forms and guide rails are set, waterproof membrane laid and side joint longitudinal tie bars fixed in position.
2. Longitudinal central joint reinforcement assembled and positioned.
3. Bottom course of concrete spread from travelling hopper.
4. Bottom course trimmed and vibrated mechanically (if required).
5. Transverse dowels inserted mechanically.
6. Mesh reinforcement manually placed onto bottom layer of concrete.
7. Top course of concrete spread by second travelling hopper.
8. Top course trimmed vibrated and screeded by second compacting machine.
9. Longitudinal surface joint groove cut in wet concrete temporary filler strip inserted.
10. Transverse joint formed with similar machine to (5).
11. Top surface given a final screeding with oscillating or vibrating finishing equipment.
12. Surface wire brushed with texturing equipment to provide skid resistance.
13. Groove filler strips replaced by flexible sealing material after concrete hardening.

PAVING EQUIPMENT

TRAVELLING SPREADER

Spreading the concrete may be performed with a blade spreader, screw spreader or box-hopper spreader (fig. 33.6). The latter is usually preferred on paving trains because of its ability to handle fairly stiff concrete mixes. It can also be loaded from the side and thereby avoid the need to bring trucks onto the paving area. The box-hopper contains approximately 3 cm of concrete separated segmentally to aid distribution. The spreading action takes place as the hopper is travelled transversely from one side of the pavement to the other during forward movement of the whole unit. The depth of the layer is set by trial and error to allow for compaction during the subsequent operations.

Data

Typical working speed	0–20 m/min
Travelling speed range	0–60 m/min
Power	20–30 kw
Width (various)	3–13 m
Operation	1 operator

Figure 33.6 Travelling box spreader

SCREEDING AND COMPACTING MACHINE

The machine comprises a row of strike-off paddles or screw blades mounted in front of the compacting beam adjusted to level the surface to the required height by pushing excess concrete forwarded to the sides. (Usually the depth set is 15–20% greater than the final slab thickness to allow for compaction.) Compaction is obtained with a transverse compacting plate, whose vibrations and amptitude may be varied by trial and error to suit the mix consistency (fig. 33.8). For most mixes the plate is commonly vibrated at from 50 to 75 Hz.

Figure 33.7 Screeding and compacting machine

Figure 33.8 Compaction mechanism

Data

Working speed	0–12 m/min
Power	20–30 kW
Paving depth	up to 350 mm
Vibrating compacting beam	50–75 Hz
Widths	2.50–12 m

DOWEL BAR INSERTION

Dowels for longitudinal and expansion joints are normally manually positioned ahead of the paving train as described earlier. However, the fixing of transverse joints dowels can usually be mechanised as shown in fig. 33.9. The unit follows immediately behind the first compacting machine releasing and vibrating dowels into the bottom course concrete.

Figure 33.9 Dowel
bar insertion machine
(transverse joints)

RAIL — DOWEL INSERTER VIBRATOR RAIL BEAM DOWEL

Data

Working speed	0–30 m/min
Frequency	60–70 Hz
Widths	2.5–12 m
Power	8–10 kW

VIBRATOR BEAM

BLADE

Figure 33.10
Transverse joint
grooving plate

JOINT FORMING

Longitudinal joint forming takes place after the top course compacting operation. A groove is mechanically cut into the wet concrete with a ploughing
tool and temporarily filled with plastic strip. Transverse joints are formed
separately using a machine almost identical to the dowel inserter unit
(fig. 33.10), but with the dowel insertions replaced by a blade spanning the
full width of the strip. The machine can also be arranged to install expansion
type joints, whereby a gap corresponding to the filler board width is first
vibrated into the concrete and then removed. The filler board and dowels are
subsequently positioned manually and vibrated down to the required depth
by the machine.

LAYING MESH REINFORCEMENT

Where a reinforced deck is required the concrete must be laid in two-course
construction. The mesh sections are stored on a wagon travelling directly
behind and to the side of the first paving train and manually placed on top
of the first course layer. Gradually, however, this operation is being updated
by a mesh-laying and depressing machine following the first paving train.

FINAL SHAPING AND FINISHING

On major roads and airfields a high quality surface without irregularities is
often specified. This can be achieved by a final finishing unit travelling behind
the joint-grooving machine. Two commonly used methods are: (i) parallel
screed and oscillating screed units and (ii) vibrating and oscillating screed
unit.

A parallel finishing screed forms part of the compactor unit and is simply
towed along so avoiding any subsequent joint-grooving operations. The beam
or blade is oscillated parallel across the line of travel at up to 70 strokes/min
with approximately 150 mm strokes. The planing angle may be varied to
accommodate different concreting consistencies and travelling speeds.

When better quality finishing is necessary an independently propelled
longitudinal finishing beam is also included in the paving train (fig. 33.11)

Figure 33.11
Oscillating finishing
beams

behind the grooving operations. The beam is suspended from a bridge and moved to and fro between the side forms at up to 15 m/min, while also being oscillated parallel to the centre line of the pavement and has the advantage of minimising disturbance to joint-groove strips.

The vibratory/oscillating finisher consists of a vibrating beam mounted just in front of an oscillating screed. The vibration effect removes high spots in the concrete and recompacts, while the oscillating beam produces the smooth surface finish (fig. 33.12). The usual configuration is the diagonal mode set at up to 60° to the line of travel to minimise disturbing filler strips (fig. 33.13). The whole unit is normally self-propelled, but when filler strip type of jointing is absent, a simple tow behind the compacting machine can be arranged (fig. 33.14).

Figure 33.12
Vibratory/oscillating
finishing equipment

Figure 33.13 Diagonal
mode for ideal finishing

Figure 33.14
Compactor-towed
finishing beam
configuration

Data

Travelling speed	0–15 m/min
Stroke	150 mm
Oscillating frequency	40–70 strokes/min
Vibration frequency	60–70 Hz
Power (self-propelled)	20 kW

TEXTURING

The type of textured finish varies depending upon designer's preference. For example, a simple brushed surface can be applied by walking a brush across the pavement or alternatively by raking or tooth combing. The equipment comprises a bridging truss to support the mechanism and/or operator.

The final task involves covering the concrete to give protection against the elements. The covers are mounted on the rails and sufficient area is necessary to allow concrete to harden sufficiently to prevent rain damage.

SAWED JOINTS

While many concrete roads are designed with wet grooved construction joints, considerable inconvenience can be avoided, for example problems of disturbing the filler strip in wet concrete by sawing the joint in hardened concrete with a diamond impregnated rotary bladed machine (fig. 33.15). Depths up to 200 mm can be cut with blades varying in thickness from 2 to 25 mm, and 250 to 500 mm in diameter. A rotation speed of 3000–4000 cycles/min is usual.

Figure 33.15 Joint
sawing machine

GUIDE   BLADE

Paving Train Data
Component parts (two-layer construction) and labour.

(i)   First concrete spreading unit (one operator).

(ii)   Screeding and compaction unit (including dowel insertion, one operator).

(iii) Mesh wagon (crane drive, truck driver, two labourers, one unloading, one operator on mesh depressing unit).

(iv)   Second concrete spreading unit (one operator).

(v)   Screeding and compacting unit (one operator).

(vi)   Jointing unit (one operator).

(vii) Finishing units (one operator).

(viii) Texturing units (from one to two operators).

(ix)   Fixing side forms (two labourers).

(x)   General labourers for tidying up.

OUTPUT RATE

Actual output $\qquad Q_a = v_p \times b \times d \times y$

Theoretical output $\qquad Q_t = v_t \times b \times d$

where b = breadth of pavement;

d = depth of concrete in the pavement;

Q = concrete quantity by volume;

y = efficiency factor depending upon the control of the units, supply of materials, gradient, crossfall and camber of the deck, etc. – approximately 0.3 to 0.8;

v = speed of the train; up to 20 m/min (V_t) can be obtained but usually 1–3 m/min (V_p) is more likely because of the problems of concrete supply.

Units are manufactured to produce pavement widths from 3 m to 12 m, assembled in modules to allow flexibility.

SLIP-FORM PAVING METHODS

The need for, and accuracy required in setting side forms for the conventional concrete paving train is time consuming. Also the constituent parts of the train demand careful control requiring a relatively high number of operators. These disadvantages have largely been removed with the development of the slip-form paving machine during the past 30 years, whereby a 'carpet' of concrete is put down without the need for semipermanent side forms. However, much stiffer mixes, with slumps of 30–40 mm are necessary for the concrete to maintain its shape without the side supports.

Several manufacturers make a variety of slip-form pavers broadly classified into: (a) track-mounted paving train with integrated side forms, and (b) track-mounted conforming plate paving machine.

INTEGRATED PAVING TRAIN

In this method of paving, the side forms are part of the machine and simply act to give shape to the concrete. The various stages in the conventional

297

Figure 33.16
Integrated slip form
paving machine

(i)

(ii)

paving train are incorporated into the unit arranged in succession from the front to the rear of the paver as follows:

(i) Spreading the concrete from a hopper or screw auger with compaction provided by built in or internal vibrating units.

(ii) Striking off the concrete to the required depth with a primary screeding beam or spreading plate.

(iii) Finishing the concrete surface with transverse oscillating float beams. Final smoothing can be given by a suspended static float plate at the rear of the paver.

Two-course paving is usual. A mesh-depressing machine is located in front of the rear paving unit, finishing beams are not required for the first-course paver. Bar insertion equipment for both longitudinal and transverse units and grooving machines can be included in the train as required.

Placer-spreader Machines

In single-course paving the concrete is often spread in windrows up to 30 m ahead of the paver to increase production. The unit is similar in construction to the paver and distributes the concrete with a screw auger, the uncompacted depth being set by means of a screeding/strike-off plate. The concrete is usually fed to the auger with a conveyor belt, which on modern machines is designed to move along the length of the auger, so ensuring uniform distribution.

Figure 33.17 Placer–spreader unit

Figure 33.18
Conforming plate paving machine

CONFORMING PLATE PAVER

The conforming plate paver is designed to produce in a single course the width and the thickness of the pavement by extruding the concrete beneath a plate mould as the unit is moved forward. Concrete in the hopper is kept topped up to a surcharge level fed by conveyor from the side and controlled with a hydraulically adjustable metering plate. In the larger versions a screw spreader is sometimes mounted just in front of this plate and acts to distribute the material more effectively. Compaction is obtained by means of a row of poker vibrators or a vibrating beam mounted across the hopper.

Transverse joint dowels can be injected with a plunger mechanism through tubes from magazines placed in front of the concrete receiving hopper. However, because of the complexity involved with a machine equipped with this sort of facility, mechanical problems have been a common feature, and as a consequence manually pre-set dowels are still quite usual.

Longitudinal joint dowels can also be mechanicaly inserted (including the temporary joint forming strip fed into a groove cut behind the conforming plate) as the paving unit moves along. Surface finishing prior to joint grooving is carried out with a transverse oscillating beam running side to side across the top of the projecting side forms. A final stationary hanging float beam is sometimes attached immediately behind this beam to give a fine finish. To avoid complications transverse joint groove forming operations are usually sawn after the concrete has hardened.

Slip Form Paver Data

Power up to	300 kW
Road widths	up to 10 m (machine width adjusted by adding inserts manually, or by built-in hydraulic equipment)
Working speed	up to 0–5 m/min depending upon the ability to supply concrete
Pavement thickness	up to 300 mm

LINE AND LEVEL

Guide wires or laser beams accurately set out each side of the paver just beyond the tracks are the common means of level control. Sensors attached to the four corners of the paving unit, either raise or lower the side forms or conforming plate.

TEXTURING AND CURING THE CONCRETE

The concrete surface must be textured to improve skid resistance and like conventional paving trains this process is commonly obtained by means of wire brushing raked either manually or mechanically propelled from a bridging unit. Good practice usually requires the finished surface to be sprayed with a curing compound and the wet concrete protected from rainfall or intense sunlight with travelling tentage. Approximately 40–60 length is necessary for 1–2 m/min travelling speed.

Production Output Example

Although paving machines are capable of travelling at 20 m/min or more, the logistical problems of supplying large quantities of concrete commonly result in no more than about 1000 m of pavement being laid during a typical day of production.

The average travelling speed of the paver being about 2 m/min. Thus for a 9 m wide road 200 mm deep and a 50% efficiency factor.

$$Q = \frac{1000}{8} \times 9 \times 0.2 \times 0.5 = 112.5 \, \text{m}^3/\text{h}$$

The time between mixing and placing should not exceed 45 min. Good practice would normally require the transport distance from batching plant to paving unit to lie with 3 km.

CHAPTER 34 SOIL-STABILISED PAVEMENT CONSTRUCTION

During the past 30 years or so, soil-stabilisation methods have been developed to increase the strength and durability of soils, to improve impermeability, to resist frost attack and reduce volume variations. The relative success in achieving these properties has led to applications in road construction, embracing the wearing surface, foundation layers and subgrade. Soil stabilisation may be achieved by:

1. Mechanically by straightforward compaction.
2. Physical reactions caused by the addition of water to cement or lime, or the solidification of hot bitumen or drying out of cut-back or emulsified bitumen, mixed with the soil.
3. Chemically by the action of two or more chemicals mixed with the soil to produce a new compound.

Figure 34.1 illustrates the application of various methods and additives.

Figure 34.1 Soil stabilisation materials

STABILISATION WITH CEMENT (RIGID PAVEMENT)

Cement-stabilised soil achieves its strength in a manner similar to concrete, in which the hydrated cement matrix bonds the soil particles together. However, unlike concrete the cement paste is not designed completely to fill the voids, but only to coat the particles, with the bonds only forming at the points of contact (fig. 34.2).

It can be seen from fig. 34.1 that cement can be used to stabilise a wide range of soils from gravel down to clay. However, coarse gravels are difficult

Figure 34.2
Stabilisation with
cement

PARTICLES
COATED WITH
CEMENT PASTE

VOID

to mix. Also fine soils are only suitable if the liquid and plastic index of the natural soil is less than 40 and 15, respectively, otherwise intensive mixing becomes difficult to achieve.

QUANTITY OF CEMENT REQUIRED

Different types of soil require varying quantities of cement depending upon the grain size and surface area, usually the finer the material the greater the cement quantity as typified in table 34.1. The range of values for a particular soil type is governed by the need to achieve strength, frost resistance, etc. The technique is suitable for producing rigid type pavement.

Table 34.1 Cement quantities and soil type

Soil type	Cement as % of the combined dry weight	7 days compressive strength N/mm^2
Sand	4–6	
Silty sand	6–10	
Silt	6–12	1–4
Clay	10–16	

Water is added during the mixing phase to bring the moisture content up to the optimum level for maximum density during compaction (Procter test). From 1 to 2% additional water should be added to allow for evaporation.

STABILISATION WITH LIME (FLEXIBLE PAVEMENT)

Lime commonly takes two forms, quicklime (CaO) and slaked lime ($CaOH_2$). Unlike cement, lime does not produce a bonded cellular structure with soil, but simply undergoes chemical changes, such that the voids are filled to form larger granules. The technique is suitable in clayey sands, silts and clays. The precise chemical reactions which take place are as yet not fully understood. In general terms the calcium oxide forms calcium hydroxide when mixed with water causing the production of heat, as a result much of the free water in the soil is evaporated leaving only that absorbed in the lattice structure of the clay component. This water is gradually drawn out by the calcium ions, removing sodium and potassium ions at the same time. As a consequence the clay begins to flocculate and subsequently coagulate to form a gel between the remaining grains of the soil and a crystalline structure is thereby obtained.

The overall effect causes an increase in the plastic limit of the soil, while the liquid limit is barely changed. Thus the 'plastic index' is reduced, with the soil becoming more granular and therefore capable of compaction with conventional equipment. Water absorption and permeability are reduced, leading to improved frost resistance. Also the resulting material can be re-worked without losing its properties, and furthermore remains stable in wet conditions. The technique is well suited for use in stabilising road subgrades or even for the road base itself. As with cement, lime-stabilised soil has water added during the admixing phase to achieve maximum possible compaction (determined by the Proctor test). If the soil to be stabilised has a greater moisture content during construction than the optimum determined from the Proctor test, then powdered quicklime is preferred, otherwise slaked lime is usually adequate.

Quantity required – 3 to 9% of the combined dry weight.

STABILISATION WITH BITUMEN OR TAR (FLEXIBLE PAVEMENT)

Bitumens and tars can be used with almost any type of soil, but for practical mixing reasons very coarse material and soils with a liquid limit and plastic index greater than 40 and 15, respectively, are usually not treatable. Unlike cement, bitumen or tar stabilisation does not rely on a chemical reaction, the strength being obtained by adhesion of the soils grains resulting from the initial tacky coating which subsequently hardens to form the bonds. After compaction a strong waterproof material is obtained. A good quality surface finish is difficult to obtain and therefore the technique is usually restricted to road base courses or minor roads. The binder is generally applied as:

(i) Emulsions of bitumen or tar.
(ii) Hot bitumen or tar.

EMULSIONS

A bitumen emulsion is designed to evaporate quickly, leaving a thin coating around the grains as shown in fig. 34.3. However, for this effect to be efficient the soil should have only 4–12% water content, very dry or very wet soil can be mixed but only with considerably difficulty.

Figure 34.3
Stabilisation with tar or bitumen

HOT BITUMEN AND TAR

Hot tar is applied at 90–100 °C, while 70–80 °C above softening point is sufficient for hot bitumen, either binder can be used with most types of soil. The best strength and compaction is obtained with soils having a continuous and proportionate reduction in grain size through the grading curve. The viscosity of the binder, should be varied to suit the grain size, for example, the larger the grain size, the more viscous the binder, and vice versa.

The main disadvantage with a hot binder is that its temperature will sharply fall on contact with the colder and wet soil particles, viscosity will increase and thus the coating process may not be satisfactory. Better results can be obtained by using an emulsifier containing dust particles. Pozzolana added in 1:1 proportion to the binder is usually sufficient. The precise quantities should be determined from test samples depending upon the specification for strength, compaction and water absorption.

BITUMEN QUANTITY

Coarse sand	4–6%	
Medium-to-fine sand	4–7%	by combined dry weight
Silt	5–8%	

STABILISATION WITH CHEMICALS (VARIOUS)

Although both cement and lime stabilisation methods involve a chemical change, chemical stabilisation usually refers to more 'esoteric' materials. These are beyond the scope of the discussion but the more widely known varieties are:

(i) Calcium and sodium chloride.
(ii) Phosphoric acid.
(iii) Polymers and resins.
(iv) Calcium acrylate.
(v) Aniline furfural.
(vi) Sulphate liquor.
(vii) (Reynolds Road Packer 233).

CONSTRUCTION OF A SOIL-STABILISED LAYER

Soil stabilisation may be carried out as follows:

1. Central mixing and subsequent laying with a paving machine.
2. Mixed-in-place with a train of machines.

CENTRAL MIXING

The soil is excavated with conventional earthmoving equipment, loaded into trucks and transported to a central mixer. Mixing plant suitable for concrete, asphalt or macadam production (according to the admixture) is commonly used depending upon the binder, filler, etc. The mix is then transported back to site transferred to a paving machine and spread in up to 300 mm deep layers. Output from 200 to 400 t/h are possible with large batching equipment.

MIXED-IN-PLACE

Mixed-in-place involves the following series of operations as illustrated in fig. 34.4.

Figure 34.4 Mixed-in-place soil stabilisation sequences

(i) Loosening the soil (top soil having been removed).
(ii) Pulverising followed by levelling and grading.
(iii) Spreading the stabiliser (and filler).
(iv) Mixing and wetting (or adding liquid binder).
(v) Compacting.

In hard compact soils, the layer to be stabilised may need to be initially loosened with a ripping or scarifying method before using a blade-type pulverisor.

Spreading

In the case of a powdered stabiliser the material is spread in about a 2 m wide carpet from a tractor or truck pulled spreader (about 5 m³ capacity), which is continuously topped up by pumping from a tanker. A liquid binder, however, is usually added during the mixing phase.

Pulverising and Mixing

Fine compact material should be thoroughly pulverised before mixing takes place. Generally the same machine can be used for both operations. Bladed versions (fig. 34.5) are suitable for soils, but more modern equipment can cold plane, pulverise and mix both blacktop and concrete pavements.

The unit comprises a segmental tungsten carbide tipped toothed rotor accommodating different cutting widths (fig. 31.13) and has an appearance similar to the road-milling machine. In contrast to blade-type machines the rotor turns counter to the travel direction and the whole unit is self-propelled. A liquid binder can be applied directly by the unit during mixing but a bitumen stabiliser is best heated to about 180 °C to avoid hardening during metering.

Figure 34.5 Pulverising and mixing unit

Mixer Unit Data

Cutting width	1.5–2 m
Mixing depth	up to 300 mm
Working speed	up to 0.5 km/h
Rotor speed	0–150 rev/min
Power	300 hp (228 kW)

Up to 300 m^2 per day can be treated.

Compaction

Final compaction is performed with convention rollers, either of the smooth wheel static type, sheep's foot or pneumatic tyred versions.

COMBINED UNITS

Machines have been developed combining all the functions of pulverising, adding the binder, mixing, spreading, levelling and compacting in a single unit, however, little information is available on their applications.

Figure 34.6 Towing units

307

BIBLIOGRAPHY: SECTION 3

1. Barber Greene Company (1979) *Bituminous Construction Handbook*. Illinois: Aurora.
2. *Bituminous Materials in Road Construction* (1962) Transport and Road Research Laboratory.
3. *The Shell Bitumen Handbook* (1990) London: Shell Petroleum.
4. Hatherly, L. W. and Leaver, P. C. (1967) *Asphaltic Road Materials*. Arnold.
5. Kezdi, A. (1979) *Stabilised Earth Roads*. Elsevier.
6. Powell, W. D. and Lister, H. W. (1978) *Compaction of bituminous materials Developments in Highway Pavement Engineering*—1. Applied Science.
7. Rogers-Martin, J. and Wallace, H. A. (1958) *Design and Construction of Asphalt Pavements*. McGraw-Hill.
8. Salter, R. J. (1972) *Highway Design and Construction*. Macmillan.
9. Scherocman, J. A. (1979) *Asphalt Pavement Construction*. American Society for Materials Testing. No. 724.
10. Sharp, R. D. (1970) *Concrete in Highway Engineering*. Pergamon.

SECTION 4   WELDING TECHNOLOGY AND BRIDGE CONSTRUCTION METHODS

WELDING TECHNOLOGY

Many aspects of construction involve the use of welded steel plate and girders. Much of this work is often carried out exposed to the vagaries of weather and the choice of welding method must therefore be carefully selected according to the conditions. Basic gas welding and manual methods are described, together with details given on the CO_2 shielded and submerged arc processes. Some of the specialised techniques such as plasma arc, laser, ultrasonic, electron beam, thermit, explosion and induction welding are briefly mentioned together with brazing. Guidelines on good practice, inspection and defects testing are also covered.

BRIDGE DECK CONSTRUCTION

In recent years new forms of bridge design and use of materials have been introduced leading to lighter, more economical solutions and increased spans. In this extensive chapter, (chapter 36), systems of bridge classification according to type of structure and method of construction are presented, the emphasis being directed towards establishing the criteria for selecting the appropriate temporary works. Specifically covered are simple beam decks, progressive placing, push-out method, stepping formwork systems, cantilever methods, stay cable and suspension bridges.

Chapter 35 WELDING TECHNOLOGY

INTRODUCTION

Welding is the joining of at least two pieces of metal by: (i) raising the temperature at the joint to melting point, or (ii) cold welding using pressure. Category (i) is referred to as 'fusion', while (ii) can be achieved in a number of ways including forge welding and resistance welding.

Fusion welding is particularly suitable for construction applications in joining metal pipes, beams, reinforcing bars, plates, fittings and fastenings, etc., in most metals including steel, cast iron, alloys and coated metals both in the ordinary atmosphere and underwater.

Methods for inspecting and testing welds without the need for destruction has also progressed to ensure that good quality is obtained and the design strength achieved. This aspect is especially important in construction work, where much of the welding uses manual methods exposed to the elements — the opportunities for automatic welding set-ups are rarely obtainable on site.

WELDING PROCESSES

FORGE WELDING

Forge welding was developed in earlier times and is the technique practised by blacksmiths whereby two pieces of metal are heated in a fire and joined by hammering. The method can be used for simple butt, tee-butt and scarf welds. Considerable craft skill is required in judging the correct temperature to achieve a satisfactory joint, and is little used in the construction industry. Modern versions of this method include hot pressure, cold pressure and friction welding which have applications in manufacturing.

RESISTANCE WELDING

Resistance welding was probably invented by E. Thompson in the 1870s. The technique relies on the heat which occurs when a current flows through a resistance, for example, the two pieces of metal to be joined (fig. 35.1). When

Figure 35.1 Resistance welding

the metal in the vicinity of the joint reaches a certain temperature the pieces can be forged together by applying pressure.

Table 35.1 Summary of welding processes

	Aluminium	Copper/alloys	Iron	Magnesium	Nickel/alloys	Precious metals	Steels	Titanium	Tungsten	Zinc	Lead
Forge	Most metals										
Resistance	✓	✓	✓	✓	✓	✓	✓	✓	✓	✓	✓
Gas (fusion)	✓	✓	✓	×	✓	✓	✓	×	×	✓	×
Shielded manual arc	✓	poor	✓	×	✓	×	✓	×	×	×	×
CO$_2$ shielded	✓	✓	✓	✓	✓	×	✓	✓	×	×	×
Submerged arc	×	×	poor	×	poor	×	✓	×	×	×	×
Plasma arc	✓	✓	✓	✓	✓	✓	✓	✓	✓	✓	✓
Laser beam											
Ultrasonic welding }	Mostly used in welding special steels such as titanium										
Electron beam											
Thermit	✓	✓	✓		✓		✓				
Explosion	Mostly used for welding dissimilar metals										
Induction	Mostly used for joining steel pipes										
Brazing	✓	✓	✓	×	✓	✓	✓	×	×	×	×

Note: ✓ = yes, × = no.

According to Joules law the heat produced J is governed by the following expression:

$$J = I^2 RT$$

where I = the current (in amps);
       R = the resistance (in ohms);
       T = the time the current is applied (in seconds);
       J = the heat produced (in Joules).

While I and T can be carefully controlled, R depends upon the pressure between the electrodes, the types of material being welded, its thickness and surface condition. The method is mostly used for welding thin plates, and so relatively

311

low voltage is required (20 V to 50 V) to produce high current from 2000 to 100 000 A. The pressure applied at the point of contact between electrodes and plates is approximately 100 N/mm^2 and the current is maintained for about half a second.

Modern equipment can be set to produce a welding time of as little as $\frac{1}{1000}$ second thus allowing the use of currents at the upper end of the range giving a consequent increase in the welding rate.

The current should only be applied when the full pressure has been obtained (Fig. 35.2) and left on only sufficient time to achieve forging heat otherwise burning of the electrodes will take place.

Figure 35.2 Resistance welding pressure–time requirements

The technique is suitable for most types of metal including, steel, copper, aluminium, zinc, lead, silver, gold and metal alloys.

Steels pose few problems for plates up to 15 mm thick, and welds have been made in 50 mm plate. Aluminium, brass and copper have lower resistances than steel and therefore require higher currents. The maximum plate thickness for these metals varies up to approximately 5 mm.

The method is widely used in the manufacturing industry and in steel fabrication workshops. However, the difficult working conditions and variety of joint shapes found on construction sites are rarely suitable for resistance welding.

Types of Resistance Welding
This method is used in many forms including spot welding, seam welding, projection welding and flash butt welding.

Spot Welding
Spot welding is commonly adopted for joining thin sheets. The equipment comprises two stem-type copper electrodes, moved from spot to spot applying current and force to each.

Seam Welding
Seam welding is similar to spot welding except that the stem electrodes are replaced by wheels. They are run quickly along in a straight line to make a 'closely-spaced' row of spot welds. This technique in effect produces a continuous weld and can be used for pressure vessels.

Projection Welding
Protrusions are pressed on to one of the sheets at predetermined locations

Figure 35.3 Spot welding

Figure 35.4 Seam welding

Figure 35.5 Projection welding

such that when the electrodes are positioned the applied force causes the weld to take place exactly at the projection and be flattened in the process.

Flash Butt Welding

Butt welding relies on the high localised heat caused by the resistance between two touching surfaces when electrical current is passed and is especially suited to welding rails, hoops, rings, etc.

The equipment consists of two electrode clamps each gripping a piece of material – one clamp is movable while the other is fixed. To obtain a simple butt weld the two pieces are brought together, the current applied and the force increased when the junction reaches sufficient temperature.

Figure 35.6 Flash butt welding

Flash butt welding is a slightly different technique for higher quality butt welds, the essential difference being that the current is switched on before the pieces are brought together. As the two parts approach, intense heat is produced and arcing takes place, such that when full pressure is applied molten metal is extruded to create a high quality joint.

Metal with Protective Coatings

If the coating will conduct electrical current, resistance welding can be achieved. Thus galvanised, electroplated, tin-plated or sheradised surfaces are all weldable, however some of the coating may be lost near the vicinity of the weld.

Non-conducting surfaces such as paint, would have to be removed before a weld could be effected.

Table 35.2 Resistance welding data

Item	Spot welding	Seam welding
Sheet thickness (mm)	0.5–15	0.5–10
Power (kW)	5–75	5–75
Welding time	0.5–10 sec/weld	60–10 mm/sec

FUSION WELDING METHODS

GAS WELDING

Gas welding is often chosen because the equipment is inexpensive and easily obtainable. Also the method is versatile and can be used for welding many different shapes and positions of joint, for a wide variety of metals up to about 25 mm thickness. Indeed the technique is an alternative to electric arc welding, the main disadvantage being that greater skill is required and the variability in the quality of the weld.

Oxyacetylene is the gas mixture commonly used as this produces 3000–3500 °C flame, however for metals with a low melting point, a lower temperature is needed to avoid the metal melting too quickly. For example a hydrogen–oxygen flame or oxygen–propane flame at about 2650 °C is more suitable for alloys.

Oxy-acetylene

Oxygen and acetylene gases react together during combustion to produce a high-temperature flame. The immediate reaction produces carbon monoxide and hydrogen:

$$C_2H_2 + O_2 = 2CO + H_2$$

The highest temperature is reached just in front of this part of the flame. Oxygen continues to react with the carbon monoxide and hydrogen as they move out into the surrounding air to produce carbon dioxide and water vapour:

$$2CO + O_2 = 2CO_2$$

$$2H_2 + O_2 = 2H_2O$$

When the acetylene and oxygen are burnt in equal proportions by volume, the flame is described as neutral. Oxidising takes place when there is an excess of oxygen. The effect can be lessened by using a deoxidising agent such as silicon with the welding rod material to form a protective layer of silica (SiO_2) on the metal plate surface. Conversely, when there is an excess of acetylene the flame is carburising or reducing, causing the metal to absorb carbon from the flame. Non-ferrous alloys and carbon steel sometimes have these requirements. For welding of most steels a neutral flame is preferred.

Shape of the oxy-acetylene flame

A neutral flame is produced by first lighting the acetylene to give a yellow smoking flame. The gas pressure is then increased until the smokiness disappears, so providing the signal to introduce the oxygen. Its pressure is subsequently raised until a sharp white cone appears at the tip of the pipe. A neutral flame (fig. 35.7a) has a clear white inner cone surrounded by a uniform and even outer light blue flame. When oxidised acetylene is present a ragged feathered white plume will be noticed around the cone within the outer flame (fig. 35.7b). Excess oxygen, however, causes a reduction in the size of the cone combined with a very streaky appearance to the outside plume (fig. 35.7c).

Gas Cylinder – Storage

Acetylene is unstable when compressed below about 1 bar (0.1 N/mm^2) and may detonate and explode. However, if the storage cylinder is filled with a

Figure 35.7 Oxy-acetylene flame settings

porous substance, such as asbestos, and the acetylene gas dissolved in acetone (which will take up many times its own volume of acetylene) the mixture is stable and made safe for transport.

Typically the gas is compressed to about $1.4 \, \text{N/mm}^2$ and stored in a strong steel cylinder.

Oxygen is also stored in a steel cylinder but compressed to about $17 \, \text{N/mm}^2$. The gas pressure at the torch must therefore be reduced, because both gases are mixed together and delivered at a common pressure, generally at from 0.01 to $0.1 \, \text{N/mm}^2$ (1–14.7 lb/in^2) depending upon the type of torch and the size of the tip. Pressure reduction takes place through regulating vales, delivered in equal proportions by volume from the mixing part of the blow pile or torch. A typical set up is shown in fig. 35.8.

Figure 35.8 Oxy-acetylene welding equipment

Blow Pipe Design

When using gas cylinders a high pressure blow pipe must be used (fig. 35.9a). However, if acetylene is produced on the premises at low pressure then it is necessary to incorporate an injector into the torch to deliver oxygen at a fairly high pressure (from 0.1 to $0.2 \, \text{N/mm}^2$) to help speed up the acetylene flow (fig. 35.9b). This type of blow pipe is known as a 'low pressure' torch and can also be used with the high pressure cylinder system if a high pressure torch is not available. However a high pressure torch *cannot* be used with the low pressure system (i.e. low pressure acetylene production) as it is unsafe to do so.

Figure 35.9 Oxy-acetylene welding blow pipe (torch)

WELDING METHODS

Assuming the welder is right-handed, the welding technique is known as either 'leftward' or 'rightward' movement.

Leftward Movement

The method is mainly used for plate under 5 mm thick and non-ferrous metals. The torch is held in the right hand and given a side-to-side motion across the joint, moving slowly forwards but trailing the filler rod held in the left hand. Hence the direction is away from the right hand (i.e. away from the body) so giving leftward movement.

Figure 35.10 Welding movements

Rightward Movement

The filler rod is trailed behind the torch thus there is less tendency for the filler material to be pushed forward and spilled on to heated parts of the plate. Also a better annealing action takes place. The method is more appropriate for plates thicker than 5 mm, up to 15 mm plate can be welded in one pass thereafter the weld should be built up in runs and layers.

Vertical Welding

Common practice adopts the leftward movement working from the base upwards (fig. 35.11). Up to 5 mm plate can be welded without any special

Figure 35.11 Vertical welding method

preparation of the joint (i.e. V-shaping). When two operators are working each side of the plate, then this limit may be increased to about 15 mm plate. Thereafter either a rightward technique, with a prepared joint working from the top downward or the bottom upwards is needed.

Overhead Welding

Either a leftward or rightward technique may be applied but considerable skill is necessary. The method relies on surface tension acting on the molten pool. The blow pipe is held almost vertically.

Figure 35.12 Overhead welding method

Figure 35.13 Weld joint design

AS A RULE OF THUMB THE ROD DIAMETER SHOULD BE ABOUT HALF THE PLATE THICKNESS

Preparation of the Joint

The configurations shown in fig. 35.13 are preferred for butt joints. The adjoining surfaces should be uniform, level and cleansed of oxides and grease.

Linde Welding

This method is often used for manual welding of pipes and tubes and uses a carburising flame. In this way any iron oxide present is reduced and the carbon absorbed, thereby providing improved fusion to give a stronger weld. Also the increase in carbon causes a lowering of the melting point and so leads to more rapid welding. The rod should contain silicon and manganese to aid deoxidisation and fluxing.

Fillet Welds

These require a slightly larger size of blow torch than for a butt weld with a similar plate thickness. Good practice requires the legs of the weld to be of equal length (fig. 35.14a).

Where the weld is to be subjected to high loading then a single V should be considered (fig. 35.14b).

For very high loads the double V is more appropriate (fig. 35.14c).

(a)

(b)

(c)

Figure 35.14 Fillet welds

Welding Rates for Steel

Plate thickness (mm)	Rate (m/h)	(One operator) Acetylene (litres/h)
2	7	100
6	3	350
10	2	500
20	0.5	1500

OTHER FORMS OF WELDING

It is not always necessary to produce a fusion weld, for example a fairly strong unfused joint can be obtained in cast-iron, steel, copper, brass, tin–bronze and aluminium–bronze by using a low melting point filler rod such as nickel–bronze, manganese–bronze or silicon–bronze. The technique requires a concentrated heat to be applied causing melting of the rod (about 900 °C) which then adheres to the plate surface. However, all surface impurities must be thoroughly removed, and a flux is generally required to assist in this process.

ALUMINIUM WELDING

Fusion welding of aluminium requires special care, as the surface coating of aluminium oxide, always present, has a higher melting point than pure aluminium. The heat should thus be applied gradually to avoid a sudden 'run' or 'holding' of the metal plate. A clean surface and use of a flux to remove the oxide is needed to reduce this effect.

319

BRAZING

Brazing is similar to soldering but uses capillary forces to cause the filler material to flow into the joint. The process involves heating the metals to a higher temperature than the brazing material by playing the blow torch over plate surface, and is thus best placed in the joint or along the joint.

Dissimilar metals can be joined in this way, including alloys. The most commonly applied brazing materials are copper, zinc, copper-alloy or silver-alloys; aluminium, however, requires alum–silicon.

The surfaces should be free from oxides, grease, etc., and a flux is normally required. The resulting joint is often as strong as the brazing metal.

FLUXES

A flux is mainly used to remove oxide from the surface of the metals to be joined and to shield the molten metal from further oxidisation. The use of a flux, however, is not universal and may not be necessary for steel, iron, lead, and copper. For most other metals and alloys, borax is the commonly applied flux except for aluminium where a special aluminium flux is necessary.

The flux can be applied as a powder or paste either directly in the joint or more commonly by dipping the welding rod.

GAS CUTTING

Metal can be severed by raising its temperature to ignition level and then applying oxygen to cause combustion and burning.

Generally an oxy-acetylene flame in the neutral setting is first applied to raise the temperature level. Pure oxygen is then introduced as a separate stream. The metal is immediately oxidised and blown away by the further heat produced from the chemical reaction.

For the cutting to be successful, the melting point of the metal must be above its combustion temperature, and any oxides present should not have a melting temperature above that of the base metal. Most steels and steel alloys possess such properties and so can be cut in this way.

Non-ferrous metals, however, commonly have resistant oxide coatings which prevent successful application of the technique.

Figure 35.15 Gas cutting torch

For applications underwater, the torch is designed to provide an air bubble surrounding the flame. This bubble is maintained by a flow of compressed air issuing around the tip of the torch.

ARC WELDING

Several different forms of electric welding have been developed in recent years including manual-arc, inert-gas shielded arc, CO_2 shielded arc and submerged arc welding. The choice of method for construction work, depends upon many factors including the opportunities to use automatic processes, weather conditions, quality of weld required, type of metal, speed, position of welding, etc.

SHIELDED MANUAL METAL ARC WELDING

The method in principle is quite simple, an electrode consisting of a metal rod acts as the filler material and is usually charged positively $(+)$ while the plate to be welded is charged negatively $(-)$. By momentarily touching the two units together the circuit is connected and current flows. On slight withdrawal of the electrode an arc is formed in the gap between plate and electrode as electrons stream across from $-$ve to $+$ve. Heat and intense light are generated in the process, the electrode and plate melted and globules of the filler rod transferred to the weld pool (fig. 35.17). The method is easier to use than oxy-acetylene welding and produces a relatively good weld. Also less skill is required.

Figure 35.16 Shielded manual arc welding

Figure 35.17 Arc welding principle

321

Electrodes

The electrode rod is made of metal appropriate for the plate to be welded, the length varying between 200 and 500 mm depending upon the diameter and the need for stiffness.

The rod is normally coated with flux which serves to produce a shield of gas as it burns (fig. 35.18), thereby partially protecting the weld from oxygen and nitrogen in the atmosphere and also to act as an ionising agent. Several compounds possess these qualities including: (i) cellulose to provide a gaseous shield; (ii) manganese to assist as a deoxidising agent; (iii) oxides of iron to control the fluidity of the slag; (iv) silicates to give strength to the covering; and (v) calcium fluoride to reduce the uptake into the weld of hydrogen from any water vapour present – hydrogen bubbles trapped in the weld lead to porosity and cracking.

Figure 35.18 Electrode rod

If excessive porosity is to be avoided the moisture content of the flux covering should thus be kept to the mininum, especially if the coating is intended to produce a low hydrogen weld. Fluorspar coating composition is best suited for this purpose.

The rods should therefore be stored in an oven and when taken out to site maintained in a protected environment until ready for use.

Welding Method

Current/voltage

Either a d.c. or a.c. power source is suitable and if mains voltage is used a transformer is required to bring the voltage down to at least 70 V for striking up the arc, with a further control down to 15 V or so to maintain the arc. The heat developed across the arc is proportional to the square of the current which should be capable of increase up to several hundred amperes depending upon the plate thickness and rod diameter.

In manual welding it is almost impossible for the welder to hold the arc length constant, as a consequence the voltage drop will change. The power supply should therefore be of the 'constant' current type (fig. 35.19) whereby variations in arc voltage produce relatively small changes in current.

Welding can be performed in almost any position but generally welding on the horizontal is preferred. As soon as the arc is struck the welder must begin to move the rod imparting a weaving side-to-side motion to ensure good fusion with the plate sides (fig. 35.20). Bad technique can cause defects as illustrated in fig. 35.21.

Figure 35.19 Typical voltage-current for arc welding

VOLTAGE (volts)

—— 'CONSTANT' VOLTAGE PERFORMANCE
- - - 'CONSTANT' CURRENT PERFORMANCE

CURRENT (amps) ——

Figure 35.20 Welding by weaving motion

→ DIRECTION OF WELDING

WEAVING MOTION

GOOD WELD

CURRENT TOO LOW

CURRENT TOO HIGH

ARC TOO SHORT

ARC TOO LONG

TRAVEL SPEED TOO SLOW

Figure 35.21 Weld defects

TRAVEL SPEED TOO FAST

For thick plate which can only be welded from one side, it is good practice to tack on a backing plate to ensure that a good root weld is obtained and to give local strength when the metal is in a soft condition.

Immediately after the run of weld has been completed the slag must be thoroughly chipped away before further runs are built up. Otherwise intrusions, holes and a generally poor weld will result.

Preparation of the Joint
The preparation of the joint should ensure that grease and oxides are cleaned away and that the edges to be joined are uniform and level with each other. For butt joints configurations shown in Table 35.3 are preferred.

323

Table 35.3 Data for butt welds

PLATE SIZE	SHAPE	Dia. OF ELECTRODES	No. OF RUNS	CURRENT (amps)
3mm	NO GAP	2 mm	1	100–120
6mm	1–6mm	4 mm	1	150–180
12mm	60° / 3mm	5 mm	2 to 3	200–250
18mm	60° / 3mm	6 mm	3 or more	300–400

(i) ROOT WELD

SINGLE V FILLET WELD

ROOT WELD
BUTT WELD

(ii)

Figure 35.22 Layering of weld runs

For fillet welds the plate thickness and corresponding rod diameter are similar to those given in fig. 35.13 for butt welds.

Welds should be built up in runs (fig. 35.22) as indicated in table 35.3.

Applications

Manual arc welding can be applied to steels and steel alloys, stainless steel, aluminium, copper, bronze, nickel and cast iron. Some may require preheating. The method is simpler to use than most other techniques.

Bronze Welding

As with oxy-acetylene welding, the manual arc method can be used to bronze-weld steel, etc. using a bronze electrode, an unfused joint will be produced.

CO_2 SHIELDED ARC WELDING

The effect of the burning flux in the manual arc welding method shields the welding zone from the surrounding atmosphere and thus reduces the oxidising reaction with the molten metal. In CO_2 welding this shield is provided by a separate stream of gas. The earliest attempts to do this simply used a permanent tungsten electrode and a surrounding jacket of inert gas, the filler wire being held manually as in the oxy-acetylene process. In modern equip-

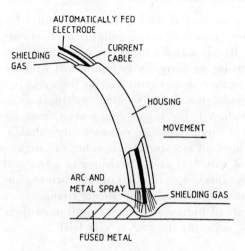

Figure 35.23 CO_2 shielded arc welding

AUTOMATICALLY FED
ELECTRODE

SHIELDING
GAS

CURRENT
CABLE

HOUSING

MOVEMENT

ARC AND
METAL SPRAY

SHIELDING GAS

FUSED METAL

ment, however, the filler wire acts as the electrode and is automatically fed through the gum as welding progresses (fig. 35.23). Unfortunately, carbon dioxide (CO_2), has a tendency to break down into carbon monoxide and oxygen at high temperatures, resulting in oxidation of the iron, and therefore must be countered with flux-covered or cored electrode. Inhert gases such as argon, called metal inhert gas (MIG) method provides an alternative to CO_2.

Welding Method

The process is generally semi-automatic using a hand gun although fully automatic equipment is available. The gas is led from a storage cylinder through flexible tubing and allowed to emerge as a stream through the shroud to shield the arc. The plate should be prepared with a V-joint as for the manual arc method and welded on the horizontal wherever possible, although vertical welds can be made.

The amount of current applied depends upon the type of metal and plate thickness as shown the table 35.4. For plate up to about 12 mm thick the globular transfer method is mostly selected (fig. 35.27). The gap between the plates should be parallel and about 5 mm if a good root weld is to be obtained. A ceramic backing tile can help to do this.

Table 35.4 Spray transfer method

Plate thickness (mm)	Electrode diameter (mm)	Current* (amps)	Travel speed (mm/min)	CO_2 gas (mph)	No. of passes
2	1.0	300	500	1	1
5	1.5	400	900	1	1
9	2.5	400	650	1	2
13	2.5	500	400	1	3
20	2.5	500	350	1	4
25	2.5	500	300	1	6

*Either a.c. or d.c. current.

The current used in CO_2 welding is significantly higher than that required in the manual-arc process, and reduces the size of the droplets or globules

325

virtually to a spray. For thick plate the wire may be permitted to touch the molten pool (dip method) causing the arc to short-circuit and thereby increase the current. In this way the tip of the wire is melted and drawn into the weld pool and the arc momentarily re-established while more wire is fed through to cause a further short circuit and so the cycle is repeated. In contrast to shielded manual metal arc welding, a 'constant' voltage power supply is generally preferred, the reason being seen from fig. 35.19 where a small change in arc length (voltage) causes a large change in current, as a consequence the burn off rate quickly alters but the arc length is restored by means of the rate of wire feed preset to deliver at a constant speed depending upon the plate thickness, etc. If a 'constant' current source is used, the voltage variation causes only very small current changes of insufficient magnitude to affect the rate of burn and the arc length must then be kept constant automatically by adjusting the rate of wire feed.

Applications
The technique can be applied to most steels, stainless steel, aluminium, magnesium, nickel, nickel–copper alloy and bronze.

The main advantage over manual arc is the continuity of welding, avoiding the need to change electrodes. The process is therefore quicker and produces less weld defects. However, when used in the open air with exposure to the elements, wind speeds exceeding about 25–30 knots tend to destroy the shielding effect of the gas and produce unsatisfactory welds. The equipment is also much more expensive.

SUBMERGED ARC WELDING

The method is generally automatic and mostly used to weld mild and low-alloy high strength steel including stainless steel and some non-ferrous alloys. Work is usually performed on the horizontal and is limited to fillet welds and long, parallel butt or seam welds. Variations in the gap dimension or irregularities lead to defects. The joint must be perfectly clean with a machined or flame cut finish.

Figure 35.24
Submerged arc welding

The Welding Method
The principles are similar to other arc-welding methods whereby an arc is struck between the tip of the electrode and plate. The filler wire forms the electrode and is fed automatically from a trolley-mounted reel as welding proceeds. Unlike CO_2 arc welding, a gas shield is not provided instead the arc is kept continuously submerged below a granular flux deposited from a

hopper as the trolley advances. In this way, fumes, spatter and intense light are virtually eliminated. Any flux not used can be collected for re-use.

The flux shields the molten pool from atmospheric oxygen and nitrogen with a covering of molten slag and produces a smooth uniformly shaped bead. Either a.c. or d.c. current may be used, but d.c. provides better control of the bead shape and welding speed, especially when a constant current type power source is used.

When low hydrogen uptake into the weld is desired (i.e. to avoid porosity) then the flux granules or powder must be dry. The flux is manufactured for specific applications and contains deoxidising compounds and sometimes substances for introducing alloys into the weld.

The joint to be welded should be prepared as for the other arc welding methods and typical welding data are given in table 35.5.

Table 35.5 Automatic submerged arc welding

Plate thickness (mm)	Electrode diameter (mm)	Current (A)	Voltage (V)	Travel speed (mm/min)	No. of passes
2	2	300	25	1800	1
5	3	450	30	1200	1
9	3	500	35	900	2
13	5	700	35	600	2
20	6	900	35	500	2
25	6	1000	35	400	2
35	6	1100	35	250	2

BACKING STRIP   TACK WELD

Figure 35.25 Backing strip and tack weld

A metal backing strip is normally required to give a sound root weld as the metal remains moulten for a long period (fig. 35.25). Also the flux hides the root of the weld and good guidance control equipment therefore is necessary.

Applications

The process permits much higher current with consequent more intense heating than other arc welding methods and is thus considerably faster. However, this can be disadvantageous causing blow holes when the variable parameters, such as fit-up, current, voltage, atmospheric conditions, travel speed, etc., cannot be accurately controlled.

The technique is mostly suited to long straight runs such as girder plates. Also large diameter pipes can be effectively welded by keeping the welding equipment fixed while rotating the pipe (fig. 35.26).

WELDING ELECTRODE   PIPE ROTATED

OUTSIDE   INSIDE

Figure 35.26 Welding large-diameter pipes

ARC CUTTING

Arc cutting can generally be carried out with identical equipment to that suitable for arc welding. The finished edge, however, is usually rough, lumpy and unsuitable for preparing parts for welding. The oxy-fuel blow torch method should be used for this circumstance.

The simplest and most readily applied cutting method is manual metal arc with a conventional metal electrode. Extra high current produces excessive heat and the weld pool becomes large and unmanageable compared to the welding mode, causing the molten metal to fall away. As in welding, the electrode is consumed. Typical cutting data are:

Plate thickness	5–10 mm
Electrode diameter	3–5 mm
Current	200–350 A
Travel speed	175–600 mm/min

SPECIALISED WELDING METHODS

In recent years the following several welding methods have been developed for specialised needs, e.g. welding special steel such as titanium, fine welds, welding dissimilar plate thicknesses, etc. Few have general application in the construction industry at present.

ELECTRON BEAM WELDING

A stream of electrons at high velocity is focused into a fine beam (approximately 0.2 mm diameter) and concentrated on the materials to be joined. If welding is allowed to take place in a vacuum, plate thicknesses up to 150 mm are possible, but this is reduced to about 50 mm in open air.

The mode of action relies on the conversion of the kinetic energy of the electrons to produce heat on striking the base metal. Because the beam is so fine, butt joints must be well finished and very accurately aligned. Most metals can be welded with this method.

LASER BEAM WELDING

A typical simple method of light amplification by stimulated emission of radiation shown in fig. 35.27 works on the principle of imparting a flash of light into a synthetic rod-shaped ruby crystal silvered at both ends. As the flash is pulsed, chromium atoms in the ruby reflect quantities of light between the mirrors cumulatively to build up a beam of light of great intensity along the axis of the rod. A small hole in the silvering at one end allows the beam to emerge. It is subsequently focused on to the target to be welded. The light beam, monochromatic and issued at the centre frequency of the spectrum, is red in colour. The resulting electromagnetic radiation directed on to the surface to be joined is designed to have sufficient concentrated energy to melt

Figure 35.27 Laser beam welding

the metal. Most metals can be welded with the technique. Laser technology is undergoing rapid development and is beyond the scope of this discussion.

EXPLOSION WELDING

Explosion welding involves placing the plates to be joined above one another separated by a gap. The bottom plate is detonated, so causing the plates to collide and weld.

Figure 35.28 Explosion welding

ULTRASONIC WELDING

Pressure is applied to the plates while very high frequency vibrations are imparted. This produces very small deformations in the plates causing frictional heat build up and subsequent welding. Melting does not take place, but welded joints as strong as the parent metal can be formed.

THERMIT WELDING

Iron oxide and aluminium powders when ignited burn at approximately 2500 °C in the following reaction:

Figure 35.29 Thermit welding

slag   steel

$$8\ Al + 3\ Fe_3\ O_4 = 4\ Al_2O_3 + 9\ Fe + heat$$

To effect a weld using these products, the groove between the plates is surrounded with a sand mould and the plates preheated to about 200 °C. The heated filler metal is then fed into the joint from a crucible causing melting of the parent material and fusion.

PLASMA WELDING

An electric arc, when constricted by argon or helium, generates temperatures in the range 10 000–25 000 °C thereby giving faster welding speeds and better quality welds compared to tungsten inhert gas (TIG) welding methods. However, the filler rod is not integral with the electrode and therefore some advantage is lost compared to the semi-automatic and automatic methods such as CO_2 shielding.

Figure 35.30 Plasma welding

INDUCTION WELDING

The method has proved feasible in welding large diameter pipes. The principle requires the butt joint to be surrounded by an electromagnetic coil to induce high local current densities and subsequent raising of temperature in the metal. Each tube is clamped and the joint pulled together. The applied pressure quickly completes the weld. High frequency current and cooling of the coil is necessary.

Figure 35.31 Induction welding

WELDING PLASTICS

Thermoplastic materials such as PVC, polystyrene, polypropylene, melt at quite low temperatures and become fluid at 100–200 °C and therefore relatively light-weight welding equipment can be used compared to that used for the welding of metals. Typical construction applications include joining sealing strip, water bar, pipes etc, using hot plates (fig. 35.32), rollers (fig. 35.33), or a welding torch (fig. 35.34). In the case of plates and rollers, the pieces of strip are simply placed above one another or lapped, or chamfered and butted together for seaming. Electrical elements heat the plates to welding temperature and pressure is applied to complete the joint.

Figure 35.32 Hot plate welding of plastics

Figure 35.33 Roller welding of plastics

Figure 35.34 Joining of plastics

Figure 35.35 Welding torch for plastics

331

For butt and fillet joints or for welding plates thicker than 2 mm the torch method is usually more applicable. Either an electrical filament (fig. 35.35b) or hot gas passing through a coil (fig. 35.35a) is used to heat a stream of air or CO_2 to to about 300 °C. Welding proceeds similar to the oxy-acetylene method using a filler rod of plastic. Typical data are given in table 35.6.

Table 35.6 Hot gas butt welding of plastics

Sheet thickness (mm)	Filler rod diameter (mm)	Travelling rates (m/h)	Joint strength as % of parent plastic
3–5	2	12–18	50–60
15–20	4	12–18	50–60

STRESS RELIEVING OF WELDS

GENERAL COMMENTS

During welding, heat causes expansion and if on contraction movement is restricted then distortions and residual stresses are likely to remain. Thus wherever possible welding should take place working from a point where the member is fairly fixed through to a free end. If this cannot be obtained then some stress relieving afterwards should be attempted, especially for pressure vessels and those parts subject to regular impact loading.

Several methods of stress relieving can be applied including furnaces and heating coils, e.g. induction heating and resistance heating. Typically for carbon steels, a re-heat temperature in the range 550–650 °C is usual, with subsequent cooling at a rate of about 150 °C per hour. Unfortunately, this elaborate procedure is virtually impossible to perform on the often large and awkward members found on construction sites and therefore good practice to ensure minimum distortion is of paramount importance.

PREHEATING OF WELDS

The common welding methods generate large temperature gradients between the local welding area and the rest of the metal. As a consequence rapid expansion and contraction with associated stressing and hardening may cause subsequent cracking and fracturing on those parts subjected to shock loading. These effects can be reduced by preheating the member in a furnace or with electrical coils but for construction work the only practical solution is generally to play a flame over the weldment area some distance on both sides of the run and bring the parts up to the preheat temperature. After welding, slow cooling or even stress relieving is desirable on critical work.

INSPECTION AND TESTING OF WELDS

The quality and therefore the strength of welds can be influenced by many factors, including the weld technique, compatibility of the metals, storage of consumables, skill of the welder, atmospheric conditions, cleanliness of the metal surface, alignment of welds, travel speed, slag intrusions (especially on multipass welds), preheating, stress relieving, etc., and not least the importance given by management to inspection and testing.

The methods used to minimise or detect faults arising from the above are broadly as follows:

1. Good practice.
2. Visual inspection.
3. Non destructive testing (NDT).
4. Destructive testing.

GOOD PRACTICE

(i)   Clean weld surfaces.
(ii)   Correct alignment and root opening.
(iii)   Parts firmly secured – usually by 10–20 mm long tack welds.
(iv)   Preheating.
(v)   Weld craters to be filled.
(vi)   Cracks to be repaired before subsequent passes laid.
(vii)   Surface slag to be removed before subsequent passes.
(viii) Broken tack welds to be removed and replaced.
(ix)   Filler material to be compatible with weld plates.
(x)   Root fusion should be complete – use a backing plate where possible.
(xi)   Position weld on the flat if possible.
(xii)   Keep consumables in the dry.
(xiii) Shield the welding area from the elements.

VISUAL INSPECTION

Check Prior to Welding

(i)   Welding procedures, specifications, documents, and appropriate techniques are clearly set out and comprehendable.
(ii)   Weld details, drawings, and sketches are clear and orderly.
(iii)   Welders are capable and experienced.
(iv)   Access platforms and working space are adequate and safe.
(v)   Weld preparation of joints, edges, etc., has been performed and surfaces are clean.
(vi)   Appropriate electrodes, fluxes, shielding gases, etc., are available.
(vii)   Dry storage facilities are available for consumables.
(viii) Tolerances, backing plates and tack welds have been properly carried out including adequate preheating.

Check During Welding

(i)   Welding equipment is performing properly and safely including current, voltage and travel speed.
(ii)   Preheating temperatures are being achieved.
(iii)   Slag surfaces are thoroughly removed after each weld run and any additional defects corrected.
(iv)   Test samples are taken.

Check After Welding

(i) Dimensions, alignments, etc., meet with requirements on the drawings.

(ii) Unacceptable defects are identified and repaired.

(iii) Cleaning and dressing of weld surfaces are properly performed and the results inspected.

(iv) Non-destructive tests are carefully executed and problems put right.

(v) Post-heat treatment provided where necessary.

NON-DESTRUCTIVE TESTING (NDT)

Non-destructive testing is applied in detecting weld defects such a flaws, fine cracks and other discontinuities which cannot be adequately detected by visual inspection. The five common methods used are:

(i) Dye-penetrant testing.

(ii) Fluorescent-penetrant inspection.

(iii) Magnetic particle testing.

(iv) Radiographic testing.

(v) Ultrasonic testing.

Dye-penetrant Testing

The surface of the weld is thoroughly cleaned with a solvent to remove dirt, grease, paint, etc. A visible dye is sprayed or poured on and sufficient time is allowed for the liquid to be drawn into cracks, holes, etc., by capillary action. The surface is then wiped clean of the dye and allowed to dry. Application of a white absorbant powder acts like 'blotting paper' to reveal the position of any surface cracks or flaws.

Fluorescent-Penetrant Inspection

A liquid fluorescent under ultra-violet light is used instead of a dye penetrant, to produce a yellowish discoloration at the positions where the liquid has penetrated into cracks, etc. The method is slightly more sensitive than the dye technique.

Magnetic Particle Testing

A current is passed between probes mounted on the length of the steel or iron part to be tested for defects. Magnetic lines of force form around a circular field in a plane at right angles to the direction of the current and any duct or crack at right angles to the field become the poles of a magnet. If the surface of the part is covered with magnetic particles, either dry or fluid-based, these will be attracted to the poles and so reveal the defect. However, only surface defects can be found in this way, and usually a machined finish is necessary for good results. Defects below the surface can be detected by passing moving search coils connected to a galvanometer over the magnetised area to detect any change in the magnetic field in the specimen.

Radiographic Testing (X-Ray Method)

A silver halide film becomes exposed when X-rays (or gamma-radiation) are directed on to its surface (fig. 35.36). However, if an object is placed between

Figure 35.36
Radiographic testing (x-ray method)

the X-ray source and film, some of the energy is absorbed by the object and a shadow produced. Thus when the film is developed, differences in the thickness of the object will be indicated by dark and light areas, i.e. thinner parts will absorb less energy than thicker parts and show up as dark areas. By careful examination of developed film, a trained observer can detect slag intrusions, gas holes and most cracks with the exception of those facing right angles to the X-rays. To determine the level of detail on the exposed film a small piece of material (the penetrameter) the same as the specimen and about 2% of its thickness is placed on the surface on the specimen during X-raying, if after developing the penetrameter is visible then defects of this size should be detectable.

Because gamma-radiation can be a health hazard, the X-ray machine must be insulated within enclosing walls of concrete or steel and therefore the specimen in general must be taken to the equipment. Where this is impracticable, special screening is necessary, with filming taking place when the workmen are away from the site.

Ultrasonic Testing

An ultrasonic probe transmitter sends very high frequency waves into the specimen which are reflected at discontinuities, such as the specimen surfaces or flaws, and converted by a receiver probe to electrical signals suitable for visual display on an oscilloscope.

In a perfect specimen only the initial pause and return surface echo would be visible, but where a flaw is present its echo will also be returned as shown in fig. 35.37. To detect all flaws in the weld it is necessary to scan along the whole length, but care must be exercised if cracks parallel to the direction of the beam are not to be missed.

Figure 35.37
Ultrasonic testing

To ascertain the size of the flaw, the probe is calibrated with a steel block pierced by a small drill hole. Test echoes indicate the amplitude of the echo from a hole of known size.

Ultrasonic testing is presently the most sensitive of all the techniques, is simple to operate, safe, and can be used on all materials. However, high levels of interpretative skill are needed.

DESTRUCTIVE TESTING

Destructive testing commonly involves the cutting out of a section of the welded plate or at least preparing samples typical of the actual weld. The common tests are:

Tensile (fig. 35.38)	Test to breaking or yield point.
Bending (fig. 35.39)	Two tests: (i) weld face in tension and tested to yield point; (ii) weld root in tension tested to fracture point.
Impact (fig. 35.40)	Brittleness tested either by the Charpy or Izod impact methods.
Hardness (fig. 35.41)	Typically involves pressing a small hardened steel ball into the surface of the specimen. The common methods use Rockwell, Vickers and Brinell equipment.
Fatigue (fig. 35.42)	Specimen subjected to alternate tension and compression.

Figure 35.38 Tensile testing

Figure 35.39 Bending test

Figure 35.40 Impact test

Figure 35.41 Hardness test

Figure 35.42 Fatigue test

UNDERWATER WELDING

In recent years there has been demand for underwater welding, particularly for offshore structures. Generally the quality of weld will not be as good as welds conducted in the normal atmosphere, mainly due to the take up of hydrogen into the molten weld metal and of course poor visibility and working conditions. In addition, the quenching effect of the water tends to produce brittleness. Conventional manual metal arc methods have proved reliable techniques to date, the electrode, however, must be waterproofed either by wrapping in tape or sealing with a coating of sodium silicate. The torch and all leads must be well insulated and connections should not come into contact with the water. Also the current must only be switched on when the electrode is in contact with the weld surface, thus good communications are necessary with the operator of the power source at the surface.

For better quality work the welder needs to work in the dry, for example in a special compressed air chamber provided with fume removal, air supply, power and communications equipment.

Chapter 36 BRIDGE CONSTRUCTION METHODS

INTRODUCTION

Bridges are a vital part of the infrastructure in most countries, linking trading and manufacturing regions, towns, etc., enabling people, goods and materials to be moved quickly and economically.

In earlier times the most successful and prolific bridge builders were the Romans. Their empire was based on trading and was well defended by the military who demanded a comprehensive network of roads and bridges. Many of the bridges constructed were made of wood and have long since disappeared, however a few examples of stonework viaducts remain in good condition throughout Europe. The technology at this time allowed only very short spans between piers, and in the case of stonework, arches and abutments (fig. 36.1) were the only feasible means of achieving strength and stability. Nevertheless, with careful siting of piers, quite wide stretches of water were successfully bridged, e.g. the River Thames in London.

Figure 36.1 Stonework arch bridge

After the collapse of the Roman Empire, European societies were overtaken by the Dark Ages, and roads decayed. Not until the setting up of turnpike trusts in Britain between 1750 and 1850 did road transport re-emerge as an important means of communication. Once again, the builders, like the Romans, turned to wood and stone for construction material and it was not until the advent of railways in the second half of the nineteenth century that serious attempts were made to use steel and iron. Rapid industrialisation during this period led to the development of stronger materials which coupled with advances in structural analysis and design allowed bridges to carry heavier loads over much wider spans than hitherto. The leading pioneers of this era included John Rennie, Thomas Telford and George Stephenson – of special note are Rennie for Waterloo Bridge, Telford for the six-hundred foot span iron-chained suspension bridge over the Menai Straits linking Anglesey to the North Wales Coast, and Robert Stephenson (George's son) for the Britannia Bridge also across the Menai Straits.

During this time advances in bridge engineering were also taking place in continental Europe and especially in the United States of America as railroads were constructed. In particular, timber and steel trusses were a feature of American developments. During the twentieth century road transport has regained its dominance and bridge technology has progressed to meet the needs of large-scale transport networks culminating in suspension bridges capable of spanning more than 600 m.

While spans have been continuously increased other developments have also been taking place, especially the use of prestressed concrete enabling much of the complex falsework and centring essential with *in situ* concrete. to be avoided. Also the change to welded cellular steel boxes assembled to form the bridge deck, has allowed more off-site fabrication compared with trusses and framed arches. As a consequence, construction times have been

Figure 36.2 Bridges classfied by structural configuration

reduced and indeed more aesthetically pleasing designs have emerged. Thus today bridges of many shapes and sizes can be seen throughout the world. Choosing the appropriate type of bridge for a particular site and function is a difficult task as many alternative solutions to a bridge problem are often possible.

Various classification systems have been devised to aid the selection process, for example structural form categorising bridges into trusses, arches, beams, slabs multipiered bridges, etc. (fig. 36.2). The objective is to try and relate type of bridge to a corresponding range of spans and heights. However, the conditions prevailing at the site so influence the choice of construction method, for example, availability of materials, ability to use floating craft, cost and skill of labour, etc., that a classification taking these factors into account would be more informative. Investigations by Wittfoht and Lampart involving many bridges constructed throughout the world indicate that the majority of modern bridges would fall into the broad categories given in table 36.1.

PREFABRICATED STEEL OR PRECAST CONCRETE BEAMS ERECTED IN PLACE OVER THE FULL SPAN

STANDARD ELEMENTS

In many countries there is sufficient demand for concrete products to enable firms to specialise in the sole manufacture of precast concrete elements. As a consequence, high production volumes are possible with a corresponding saving in costs. Bridge engineers have taken advantage of this market by specifying designs which utilise standard beams of relatively short span, 10–40 m being the typical manufactured range.

PRECAST BEAM SECTIONS

The common sections now available are:

(i) I-beam.
(ii) Inverted T-beam.
(iii) Hollow box beam.
(iv) M-beam.

In general the units are prestressed and pretensioned on the long line process, although ordinary reinforced members are not precluded.

I-Beams
Normally these are simply supported between the bridge piers and individually separated by evenly spaced transverse diaphragms. The deck is formed *in situ* and thus generally requires extensive formwork.

Inverted T-Beams
These are butted together and act compositely with concrete between and over the units, formwork only being required to form parapets, kerbs, etc.

Table 36.1 Common bridge construction methods

Type of bridge according to method of construction	Span (m)	Length of bridge (m)
1. Prefabricated steel or precast concrete beams lifted into place over the full span		
(i) standard beams	10–40	
(ii) non-standard beams	40–250	No limit
(iii) steel trusses	50–100	
2. Prefabricated steel or precast concrete beams progressively placed by gantry girder method	30–60	300–600
3. Precast concrete or prefabricated steel deck pushed out (constant radius deck only)	25–60	300–1200
4. Prefabricated or precast elements or *in situ* concrete structures supported by falsework founded directly on the ground	5–125	5–125
5. *In situ* concrete deck constructed on stepping formwork	20–60	300–600
6. *In situ* or precast concrete segments constructed in cantilever and progressively post-tensioned	50–150	No limit
7. Prefabricated steel or precast concrete segments stayed by cables	75–800	No limit
8. Prefabricated steel or precast concrete segments suspended from cables	Span between towers 400–1500	No limit

Note: Values are typical ranges. Where long bridges are involved, many spans may be necessary, and use of simple short span systems may be more economical than a few long spans.

BOX

I-SECTION

150 0

M-BEAMS WITH TOP AND BOTTOM SLAB

1000 1000

LUGS REINFORCED IN SITU SLAB

CROSS TIES

I-BEAM IN STEEL

Figure 36.3 Precast beam sections

M-Beams

Inverted T-beams are used with permanent formwork to cast the slab and produce a voided composite structure.

Hollow Box Beams

The beams are butted against each other and transversely post-tensioned. The deck is added *in situ* and usually is not designed to act compositely with the beams.

Steel Beams

Steel continues to feature in simply supported spans using I-sections. The choice of material largely depending upon the relative costs of steel and concrete. Like concrete elements the slab is often designed to act compositely witht the beams and therefore extensive temporary formwork is a requirement.

HANDLING PRESTRESSED CONCRETE ELEMENTS

Figure 36.4 Lifting beam for precast units

Pretensioned single-span members are designed as simply supported spans and usually cannot act as cantilevers. Thus units should always be lifted at the hooks cast in place by the manufacturers. Where these have not been provided then the lifting point of slings should be near the ends. A lifting beam would be preferable (fig. 36.4). Beams may be stacked for temporary storage, with timber spacing blocks located near the ends and positioned vertically above one another (fig. 36.5).

CORRECT   INCORRECT

Figure 36.5 Stacking arrangement of pretensioned units

BEAM ERECTING

Placing standard beams between abutments is usually quite straight forward using conventional cranes, such as crawler or truck-mounted boom types. The units are generally brought in by trailer and temporarily stored near the crane. The crane itself can be positioned between the abutments to suit the achievable lifting radius. Alternatively, for a rail or river crossing or when difficult ground is encountered, the beams may be placed from the bridge approach road (fig. 36.6). However, a large crane would be required to provide lifting capacity at about 20–30 m radius and methods would thus for economical reasons be generally limited to light steel beam bridges.

COMPOSITE CONSTRUCTION

In composite design the deck acts as the compression member while the beams largely serve to take tension. During construction the deck concrete is placed

Figure 36.6 Beams erected by crane

in situ and therefore must be both supported and allowance made for deflection of the beams, if an unsightly line along the bridge is to be avoided.

The problem is particularly acute with steel beams which deflect markedly under load and therefore a camber is normally induced before concreting takes place. The simplest method requires a frame slung under the deck with a post located below the centre point of each beam (fig. 36.7). Screw or hydraulic jacks are used to put in the desired camber and released when the concrete has reached the required strength. The frame, constructed as a truss with a working platform to aid both the jacking and shuttering operations, is hung from the bottom flange of the two beams suspended from a roller device to aid transfer from one span to the next.

Figure 36.7 Composite construction method

NON-STANDARD BEAMS

Many bridges are designed to achieve particular aesthetic lines, for which relatively short, standard concrete beams are generally inappropriate, whereas purpose-made units can be formed to shape as required. Although the beams for such bridges would be much heavier than standard units, modern developments in cranage have made lifting by this method quite feasible. For example, strut boom truck-mounted cranes are available with capacities over 1200 tonnes and the more versatile hydraulic-boom truck-mounted cranes are currently capable of lifting over 800 tonnes, but these are expensive to hire and set up.

For work over water, crane-mounted pontoons and shear legs with capacities over 2000 tonnes can sometimes prove more economical in long multi-span bridges where several sections are being constructed simultaneously (fig.

Figure 36.8 Beam erection by crane and/or pontoon

Figure 36.9 (a), (b) and (c) Specialised lifting of large and heavy units

36.8). Unfortunately, however, strong currents and tidal waves often present severe difficulties in both the launching and transporting operations. Also a suitable prefabrication area and loading wharf are generally required and therefore bridges through step valleys are normally unsuited to this mode of construction, although with several minor bridges it has proved viable to prefabricate units at a distant yard and tow the members several miles to the bridge location. For large spans involving heavy pre-assembled trusses or prefabricated box sections beyond the capacity of normal cranage, specialised lifting methods are generally required, for example, box units weighing over 2500 tonnes and 300 m long have been successfully raised from temporary bunds (fig. 36.9). Where this would not be possible, some form of temporary support must be provided at the piers (fig. 36.10).

The method facilitates fabrication under factory conditions, away from the site, giving the potential for improved quality control and productivity. Towing to position can be planned to coincide with favourable weather windows with the minimum erection period exposed to the vagaries of the elements.

(i)

(ii)

Figure 36.10 Barge transport of heavy units

BEAMS PROGRESSIVELY PLACED BY A GANTRY GIRDER

Where many regular simply supported spans predominate, for example long bridges of 600 m or more, then construction of the deck working out from the abutments using a gantry girder is sometimes more economical than using conventional cranage or pontoons.

The gantry technique is suitable for spans of 30–60 m involving beam units up to 250 tonnes, with the gantry girder itself often weighing up to 150 tonnes or more for beams of this size.

Wherever practicable, the casting or fabricating yard(s) should be set up on the approaches to the bridge to minimise transport problems. Typically the previously placed beams are used to support the transport bogies. Figure 36.11 shows three views of a commonly adopted procedure. The beams are loaded on to two wagons and then transferred to either bogies or hangers running on rails along the previously erected deck. The gantry then distributes the units into place between the piers. Rails temporarily fixed across the deck and on the far pier are provided for traversing the gantry. When all units are in position, the girder is winched forward on rail-mounted bogies over the next span. Thus the gantry needs to be slightly longer than twice the span.

Finishing of the roadway deck follows up behind the units.

Figure 36.11
Progressive placing of precast concrete/steel beam by gantry method

INCREMENTAL LAUNCHING OR PUSH-OUT METHOD

In this form of construction the deck is pushed across the span with hydraulic rams or winched (fig. 36.12). Decks of prestressed post-tensioned precast segments, steel or box girders have been erected. Usually spans are limited

346

Figure 36.12
Incremental
launching/push-out
method

to 50–60 m to avoid excessive deflection and cantilever stresses, although greater distances have been bridged by installing temporary support towers (fig. 36.13). Typically the method is most appropriate for long, multispan bridges in the range 300–600 m, but, much shorter and longer bridges have been constructed. Unfortunately, this very economical mode of construction can only be applied when both the horizontal and vertical alignments of the deck are perfectly straight, or alternatively of constant radius. Where pushing involves a small downward grade (4–5%) then a braking system should be installed to prevent the deck slipping away uncontrolled and heavy bracing is thus needed at the restraining piers.

Bridge launching demands very careful surveying and setting out with continuous and precise checks made of deck deflections. A light aluminium or steel launching nose forms the head of the deck to provide guidance over

Figure 36.13
Temporary support for
push-out method

the pier. Special teflon or chrome-nickel steel plate bearings are used to reduce sliding friction to about 5% of the weight, thus slender piers would normally be supplemented with braced columns to avoid cracking and other damage. These columns would generally also support the temporary friction bearings and help steer the nose.

In the case of precast construction, ideally segments should be cast on beds near the abutments and transferred by rail to the post-tensioning bed, the actual transport distance obviously being kept to the minimum. Usually a segment is cast against the face of the previously concreted unit to ensure a good fit when finally glued in place with an apoxy resin. If this procedure is not adopted, gaps of approximately 500 mm should be left between segments with the reinforcement running through and subsequently filled with concrete before post-tensioning begins. Generally all the segments are stressed together to form a complete unit, but when access or space on the embankment is at a premium it may be necessary to launch the deck intermittently to allow sections to be added progressively. The corresponding prestressing arrangements, both for the temporary and permanent conditions would be more complicated and careful calculations needed at all positions.

The principle advantage of the bridge-launching technique is the saving in falsework, especially for high decks. Segments can also be fabricated or precast in a protected environment using highly productive equipment.

For concrete segments, typically two segments are layed each week (usually 10–30 m in length and perhaps 300 to 400 tonnes in weight) and after post-tensioning incrementally launched at about 20 m per day depending upon the winching/jacking equipment.

CONSTRUCTION WITH FALSEWORK

BEAM DECKS

While bridges designed for standardisation or regularity are often more economically built with precast concrete or prefabricated steel sections other considerations such as aesthetics may be overriding requirements. In cases where beam depth varies and geometry is complicated, or pier design is awkward, the most feasible construction solution generally involves falsework to support either an *in situ* concrete or a precast concrete deck, or combination of both (fig. 36.14). Bridges built at heights of 30–40 m above ground are practicable and spans over 100 m are not uncommon.

Not withstanding these factors, much simpler bridges, such as those normally considered appropriate for construction with standard precast

Figure 36.14 Complex falsework for *in situ* concrete bridges construction elements (beams) are frequently constructed *in situ* because the falsework and formwork would involve little more than scaffold tubes and timber or proprietary shuttering systems, so enabling a local contractor to carry out the work (fig. 36.15). Furthermore, falsework supports avoid the need to design the permanent structure to carry imposed loads generated only during the construction phase.

Figure 36.15 Standard falsework for *in situ* concrete deck

Unfortunately the technique is labour-intensive and wasteful of shuttering unless many re-uses can be obtained. Also great care has to be taken in preparing the foundations for supports, checking that members are plumb, timbers well seated and joints tight before concreting begins. There is also more likelihood that an *in situ* concrete finish will vary because of the difficulties of quality control on site. For these reasons precast elements supported on falsework and subsequently post-tensioned have been commonly chosen as an alternative to *in situ* work (fig. 36.16). Prefabricated box segments have proved suitable for multispan bridges, up to about 40 m span. Over 100 m of completed deck per week can be achieved with this method. A further advantage lies in the ability to drape the prestressing cables to give the most efficient use, and the post-tensioning is performed for each span compared with the stage stressing required for the alternative of cantilever construction.

Figure 36.16 Beam falsework for precast post-tensioned concrete deck

CONCRETE ARCHES

Concrete arches have been popular because of their inherent pleasing aesthetics and the shape is also ideal to take up compressive stress. Unfortunately, however, complicated falsework is usually required, whether the construction is in precast or *in situ* concrete. A common construction procedure involves driving temporary piles into the river bed to act as foundations for the falsework. Depending upon the weight to be carried the support system is made up of single scaffold tubes and timber beams, or specially designed struts and Warren girders for heavier loads.

A complete soffit of formwork is prepared between the abutments (fig. 36.17) and concreting of the deck carried out in the sequence shown in fig.

349

Figure 36.17
Falsework for *in situ*
concrete arch

Figure 36.18 Pour
sequence for *in situ*
concrete arch

Figure 36.19
Falsework for deck
supported off completed
arch

Figure 36.20 'Hanging'
falsework/formwork for
arch construction

36.18 thereby keeping movement of the falsework to the minimum. After completion of the arch the piers and general superstructure are constructed either conventionally with the aid of falsework (fig. 36.19) or with precast units placed in position by a crane mounted barge.

In deep ravines it is often not feasible to support the arch formwork from piled foundations, because of the depth and awkwardness of the site. A common solution has been to hang the falsework from the completed abutments as shown in fig. 36.20 enabling the arch to be progressively concreted section by section, with perhaps the middle third built on centring assembled at the bottom of the valley and hoisted into place (fig. 36.21).

350

Figure 36.21
Construction of centre
arch span

Figure 36.22 Ring
compression induced in
arch configuration

Figure 36.23
Falsework for precast
concrete arch

During the period of popularity of slender arch bridges built of *in situ* reinforced concrete, several major bridges were also designed as precast construction. By utilising ring compression (fig. 36.22) the complete arch can be assembled from independent precast hollow boxes, which in the final condition behave similarly to the traditional stone arch, thus requiring no monolithic connection between sections.

A typical construction arrangement is shown in fig. 36.23, where it can be seen that a substantial falsework structure is needed to support the precast units. Each unit is brought to the winching mast at the centre of the span and raised to the crown of the arch (fig. 36.24). Units are subsequently lowered on bogies to the abutments, gradually building up equally to the crown. Flat jacks are incorporated in the joints at the quarterpoints and after insertion of the final unit, the structure is raised into a freestanding condition above the falsework and the joints concreted.

The superstructure is subsequently added *in situ* or as precast concrete elements.

STEPPING FORMWORK CONSTRUCTION

While falsework standing on the ground is often viable for a single span or at most a few spans, the use of stepping formwork usually proves more econ-

Figure 36.24 Typical erection system for precast concrete arch segments

omic for long bridges of 300 m or more. Individual spans between piers generally lie between 20 to 60 m, and the deck dimensions should be reasonably constant to avoid expensive formwork adjustments. This method is appropriate where a bridge is constructed over poor soil, water, or is high above ground level. Like cantilever construction, standardisd formwork and equipment is used many times over, thereby avoiding the need for almost continuous re-learning as is often the case with traditional bridge-building techniques, such as steel frames, concrete arches, formwork and falsework systems, etc.

The *in situ* deck is supported on movable girders temporarily carried by the bridge columns or piers. After sufficient time has elapsed for the concrete to gain strength, the individual units are moved forward to the next span. Where possible the formwork is permanently attached to the girders and arranged to fold down during the stepping phase. However, this depends upon the shape of the cross section of the bridge and with more complicated designs, stripping and re-assembling the shutter panels may be unavoidable.

Basically two methods have proved successful: namely with or without auxiliary girders.

Figure 36.25 Stepping formwork system for *in situ* concrete deck

AUXILIARY GIRDERS

The stepping girders are made a little longer than the casting span, with the service or auxiliary girder required only during the stepping operation. In the example shown, the forward ends of the stepping girders are supported on a collar attached to the pier or column, while the rear ends are carried on a beam hung from a cross member seated on the completed deck. After concreting, the girders are moved forward singly or in pairs, by attaching the front part to travelling bogies mounted on the auxiliary girder, while the rear is hung from rail-mounted bogies travelling on the newly completed deck. The auxiliary girder being slightly longer than two span lengths is then moved forward to bridge the next span and the cycle repeated.

WITHOUT AUXILIARY GIRDERS

The stepping girders are approximately two times the span length, with the centre section made strong enough to carry the temporary loads and deck during construction (fig. 36.26). The 'flying' sections are only needed during the stepping phase, and therefore can be of light fabrication, usually in the form of lattice beams. The complete unit is carried on collars fixed temporarily to the columns. Units are moved forward either singly or in pairs. As soon as the front section of the girder reaches the column support, the rear collar can be released and transferred to the columns on the next span. A small crane mounted on the forward section of the girders is normally used to facilitate this operation. A separate crane travelling on the completed deck

Figure 36.26 Stepping formwork system without auxiliary girders

is usually selected for concrete placing and handling of materials, equipment and shuttering. A horizontal boom crane is generally the most suitable type in order that the carriage can be kept well clear of the working area of the deck. Bridges curved on plan with a radius down to about 400 m have been successfully completed. For the method to be economic it is necessary to obtain many re-uses of the formwork and falsework and therefore a divided carriageway is often favoured.

Typically 300–400 m² of completed deck surface area can be achieved in a week.

HANGING METHOD

The use of stepping girders mounted underneath the deck tend to lose their advantage when designs involve complicated arrangements at the column or pier heads, e.g. mushroom shapes, haunches, etc. To deal with such problems, bridge builders have occasionally used the hanging method with stepping girders. A typical example is shown in fig. 36.27. The equipment comprises an integrated unit of formwork, yokes and girders. The sequence of operations requires the rear of the stepping girder to be supported on the previously constructed deck, the other support being provided at the next pier on a temporary base which is subsequently removed and concreted into the deck. The centre beam of the whole unit extends over two full spans to facilitate the stepping action.

The section of deck at the column head is concreted first, followed by the mid span. Two side supports are then temporarily positioned on the completed deck, to relieve the forward support, which is then transferred to the far end and secured to the next column. The whole unit is then winched forward a full span and the cycle repeated. The formwork must be designed to hinge away from the concrete deck, and the central portion, obstructed by

Figure 36.27 Stepping formwork-hanging method

the column, should be capable of folding back during travelling. Output rates similar to the previously described girder methods have been achieved, i.e. 350–400 m² of deck surface area per week, stepping units weighing over 500 tonnes are typical for spans up to 40 m.

CANTILEVER CONSTRUCTION WITH PRESTRESSED CONCRETE

The erection of bridges span-by-span, working out from the abutments towards the centre, has proved very economical as witnessed by the popular use of stepping formwork or bridge-launching techniques with problems such as difficult terrain, and transport of materials made much easier. The advent of cantilever construction methods, during the past 25 years now allows many of these advantages to be extended to bridges with spans in the range 50–150 m. The method can be applied to *in situ* concrete, precast concrete or steel decks, and is normally carried out by balanced cantilever or occasionally by progressive placing methods.

BALANCED CANTILEVER CONSTRUCTION

Figure 36.28 Typical box section configurations

Developments in box section (fig. 36.28) and prestressed concrete led to short segments being assembled or cast in place on falsework to form a beam of full roadway width (fig. 36.29). Subsequently the method was refined virtually to eliminate the falsework by using a previously constructed section

Figure 36.29 Concrete
box section/bridge deck

of the beam to provide the fixing for a subsequently cantilevered section. The principal is demonstrated step-by-step in the example shown in fig. 36.30.

In the simple case illustrated, the bridge consists of three spans in the ratio 1:1:2. First the abutments and piers are constructed independently from the bridge superstructure. The segment immediately above each pier is then either cast *in situ* or placed as a precast unit. The deck is subsequently formed by adding sections asymetrically either side.

Ideally sections either side should be placed simultaneously but this is usually impracticable and some imbalance will result form the extra segment weight, wind forces construction plant and materials (fig. 36.31). When the cantilever has reached both the abutment and centre span, work can begin from the other pier, and the remainder of the deck completed in a similar manner. Finally the two individual cantilevers are linked at the centre by a key segment to form a single span. The key is normally cast *in situ*. Typical changes in bending moment during construction are shown in fig. 36.32.

Figure 36.30 Balanced
cantilever construction

Figure 36.31 Effect of
equipment materials
and wind

Figure 36.32
Changes in bending
moment by adding units

EXAMPLE: INCREASE IN BENDING MOMENT
ALONG CANTILEVER WHEN SECTIONS ADDED

PERMANENT PIER

Figure 36.33
Counterbalance/support
and anchor arrangements

COUNTERBALANCE

SUPPORT

SUPPORT

ANCHOR

In cases where the span between the abutment and the pier is shorter than that reached out to the centre, the overturning moment must be resisted by a downward reaction, i.e. balast or an anchor (fig. 36.33) at the abutment end.

Application of Prestress
The procedure initially requires the first sections above the column and perhaps one or two each side to be erected conventionally either in *in situ* concrete or precast and temporarily supported while steel tendons are threaded and post-tensioned (fig. 36.34). Subsequent pairs of sections are added and held in place by post-tensioning (fig. 36.35) followed by grouting of the ducts. During this phase only the cantilever tendons in the upper flange

FIRST SECTIONS SUPPORTED
ON FALSEWORK

TEMPORARY
FALSEWORK

Figure 36.34
Temporary falsework for
sections over piers

Figure 36.35 Precast
concrete sections added
in pairs and post-
tensioned

ANCHORED
CONCRETE SEGMENT

357

Figure 36.36 Typical tendon and duct configuration

and webs are tensioned. Continuity tendons are stressed after the key section has been cast in place (fig. 36.36).

The prestressing system would normally be one of those described in chapter 30 thus the final gap left between the two half spans should be wide enough to enable the jacking equipment to be inserted. When the individual cantilevers are complete and the key section inserted the continuity tendons are anchored symmetrically about the centre of the span and serve to resist superimposed loads, live loads, redistribution of dead loads and cantilever prestressing forces (fig. 36.37).

Figure 36.37 Continuity tendons

The Advantages of Continuity

The earlier bridges were designed on the free cantilever principle with an expansion joint incorporated at the centre. Unfortunately, settlements, deformations, concrete creep and prestress relaxation tended to produce deflections in each half span (fig. 36.38), disfiguring the general appearance of the bridge and causing discomfort to drivers. These effects coupled with the difficulties in designing a suitable joint (fig. 36.39) led designers to choose a continuous connection, resulting in a more uniform distribution of the loads (fig. 36.40) and reduced deflection. The natural movements were provided for at the bridge abutments using sliding bearings or in the case of long multi-span bridges, joints at about 500 m centres (fig. 36.41).

Figure 36.38 Deflection and distortion effect

Figure 36.39 Centre hinged joint

Figure 36.40 Centre continuous joint

FREE CANTILEVER

BENDING MOMENT

Y > X

CONTINUOUS SPAN

JOINTS FOR HORIZONTAL MOVEMENT

APPROX 500m

Figure 36.41 Joints for horizontal movement

CONSTRUCTION SYSTEMS

Spans up to 200 m are commonly built using the balanced cantilever method, thereafter the dimensions of the box section increase quite rapidly, becoming very heavy and temporary supports impracticable. Congestion of prestressing tendons is also a problem. Four systems of construction are commonly used as follows:

1. Travelling former and cradle (fig. 36.42).
2. Travelling former with overhead gantry girder (fig. 36.43).

Figure 36.42 Travelling former and cradle for *in situ* concrete balanced cantilever method

FIXED TO PREVIOUS SECTION

MOBILE CARRIAGE

CRADLE TO CARRY FORMWORK

FORMWORK

Figure 36.43 Travelling former with overhead gantry for *in situ* concrete balanced cantilever method

Figure 36.44 Precast segments erected with mobile hoist

Figure 36.45 Precast segments supported with overhead gantry

3. Precast segments with mobile hoist (fig. 36.44).
4. Precast segments with overhead gantry girder (fig. 36.45).

Precast construction is generally much faster than the *in situ* alternative, but is only appropriate where precasting facilities can be conveniently and economically arranged. The method also tends to be a little less labour-intensive and would be favoured unless the weight of units becomes excessive and difficult to transport or launch and lift into position. Control of the bridge alignment is protracted with precasting, whereas the *in situ* method facilitates differences between actual and predicated camber to be more easily controlled during construction.

Travelling Former and Cradle

The equipment comprises adjustable forms carried by a steel cantilevering truss and cradle temporarily anchored to the previously concreted section and mounted on rail track positioned above the webs (fig. 36.46). The sequence of operations typified in fig. 36.47 is as follows:

Figure 36.46
Travelling former and cradle

Figure 36.47
Concreting sequences
with travelling former

1. Post-tension previous section, strike the formwork and move carriage.
2. Position outside form and inside flange form, fix reinforcement, ducts and cables and concrete bottom flange.
3. Fix web reinforcement, position inside formwork, fix top flange, reinforcement, ducts and cables and concrete webs and top flange.
4. Cure concrete.
5. Repeat cycle.

Crane assistance, while not essential, often proves beneficial for concrete placing and materials handling operations. An efficiently operating cycle takes approximately 7 days to complete. Thus with two 3 m long carriages operating, about 1 m per day of advance is possible. Attempts have been made to speed up this rate, for example by improving the curing period, and increasing the length of the segments. This latter method has generally proved too costly and the most economic section length seems to be 2–4 m. Other alternatives involve precasting the webs or arranging for the deck design to allow the top flange to be completed after post-tensioning of the section. A rate of advance of 2 m per day for a pair of carriages has been obtained in this manner.

Travelling Former with Overhead Gantry
Improvements in the rate of advance over cradle methods can be achieved by having larger sections carried by a gantry girder bridging a full span. However, the cost of the girder is high and only bridges 500 m or more in length can be competitively constructed in this way. Sections up to 10 m long are possible but the spans should be limited to less than 150 m, thereafter the girder itself rapidly increases in size, weight and capacity. The use of the gantry girder offers advantages with respect to materials handling, whereby water or ground level transport can be largely avoided for at least the deck construction phase.

Figure 36.49 illustrates a typical sequence of operations. The system comprises a main girder approximately one and a half spans in length carrying two travelling cradles. Construction of the deck section proceeds

Figure 36.48
Travelling former and
overhead gantry (cross
section)

Figure 36.49
Travelling former and
overhead gantry
sequences

symmetrically from each column or pier head. The girder is supported in the
working position on pads placed on the top flange of the deck at the web
positions above each pier head (step 1). When each half span is complete and
keyed together the main girder is travelled forward and temporarily propped
at the next pier (step 2). The section immediately above this pier is then
constructed and subsequently used to provide support for the girder during
projection to the next half span point (step 3). The travelling cradles are
thereafter repositioned and the cycle of operations repeated. It can be seen
in the figure that the lower flange formwork is hinged to allow passage past
the pier section.

The period of the construction cycle is similar to that for the independent
system and approximately 20 m of advance per week is possible using 10 m
long cradles.

Precast Segments Erected with a Mobile Hoist
The use of precast segments has several advantages over *in situ* concrete, for
example, better quality control, manufacturing rates not dependent upon the
erection programme, simpler erection methods and equipment.

The general concept, as with *in situ*, is to attach the segments in pairs
to maintain symmetry either side of the pier, followed by immediate post-
tensioning and grouting of the ducts. The segments can be delivered and
placed in position by several methods, crane handling being most common
where easy land access is available (fig. 36.50). Occasionally a crane-mounted

Figure 36.50 Precast
segments supported by
needle beams

Figure 36.51 Needle
beam equipment and
post-tensioning
sequences

Figure 36.51 Needle
beam equipment and
post-tensioning
sequences

barge may be used over water, but this tends to be too troublesome for most
situations, as a consequence a deck-mounted lifting device is generally
preferred (fig. 36.51). Such equipment is relatively simple and typically
comprises a mobile hoist mounted on two beams and anchored to the
previously erected segment and projecting beyond it. The segment is raised
from a barge or trailer below, moved into position, and working from a
hanging platform post-tensioned as quickly as possible (in the case of erection
by crane, a temporary tie beam placed across the top flange to the completed
deck is usually necessary to maintain steadiness and support the working
platform).

The precast units are manufactured with matching joints (see chapter 30)
and given a thin (1 mm) coating of epoxy resin glue prior to post-tensioning.
However, since full shear strength does not develop immediately (and reinforcing steel is not continued through as with the *in situ* system) shear keys
are normally required.

Units 2–4 m long and weighing up to 60 tonnes are commonly erected
with up to four segments put in place each day (about 40 m per week on
average). A deck area of at least 5000 m^2 is necessary to justify the costs of
the equipment.

Precast Segments Erected with A Launching Girder

The advantage of using a launching gantry largely depends on the ease of
handling materials, for example to avoid the problems of water transport by
bringing the units along the previously constructed deck. Unfortunately,
however, when the span exceeds 150 m or so, the weight of the deck segments
require a very heavy girder, and the method then tends to become very
rapidly uneconomic. Also, unless the bridge is multispan with a deck area of
at least 10,000 m^2, other bridge types would probably be favoured.

There are two categories of launching gantry:

1. Gantry slightly longer than a full span.
2. Gantry slightly longer than twice the span.

Figure 36.52 illustrates the first type, with the gantry supported on two
trolley-mounted legs, ones rests on the previously constructed cantilevering
half span, A, while the other is set on the segment above the next pier, B.
Segments are introduced along A, tuned through 90° on the girder carriage,
positioned and post-tensioned (step 1). On completion of two complete half
spans the keying section is inserted to complete a full span and the gantry
moved forward until the trailing leg C can be set down on the temporary

Figure 36.52 Precast segments erected with gantry slightly larger than full span

bracket of the next pier (step 2). The pier segment is then placed, a temporary support, D, erected and the final stage of the gantry manoeuvre completed (step 3). The weight of the gantry is transferred from D onto B and the next cycle of segment placing commenced. (Figure 36.53 illustrates an application of a combined design of the B and D support). Figure 36.54 shows the second

Figure 36.53 Interchangeable gantry support legs

Figure 36.54 Precast segments erected with gantry slightly longer than twice full span type of gantry, which allows the support to be positioned above the piers during the segment erecting phase, thereby facilitating improved stability.

Gantry girders are typically used for placing segments weighing up to 100 tonnes but examples of over 250 tonnes segments exist. Rates of advance of 40 m per week (that is, four segments per day) are fairly typical. A point to consider when planning balanced cantilever construction is the construction of the segments immediately above the piers, especially when using conventional falsework as several weeks must be allowed for this phase.

CONTROL OF STABILITY DURING CONSTRUCTION

Disequilibrium in weight of each half span and the loads carried during construction, can be stabilised by the following methods:
1. Making a rigid connection between the deck and piers.
2. Temporarily tie the deck to the piers (fig. 36.55). Concrete wedges are used as the bearing pads and subsequently removed with the aid of the flat jacks and replaced by the permanent bearings.
3. In the case of very slender piers provide additional stiffening and support.

Figure 36.55
Temporary ties for balanced cantilever method

CONTROL OF DEFLECTION AND CAMBER

During construction prestress losses caused by steel relaxation, creep and shrinkage in the concrete, service loads and self-weight, non-uniform prestress, temperature. etc., cause the arm of each cantilever to deflect and distort the design profile. Compensating cambers should therefore be incorporated by making adjustments of the formwork moulds or precast segments. In particular, good quality control of the concrete is essential in order that the creep and strain can be minimised. For a detailed treatment of methods for calculating the camber adjustments see Mathivat, reference 12, Bibliography: Section 4.

PRECASTING SYSTEMS

The two principle methods are the short line and long line systems – both adopt the principle of match casting, such that the face of the previously concreted section acts as the form for its next adjoining segment. The resulting joint should be a perfect fit if the segments are then erected in the same sequence.

Short Line Method

A single set of shutters is used in conjunction with a permanent soffit (fig. 36.56) to cast a single segment. The sequence of operations is as follows (fig. 36.57).

1. Cast new segment against the old segment and steam cure overnight.
2. Take old segment to storage.
3. Release formwork from new segment.

Figure 36.56 Formwork system for short line segment casting

Figure 36.57 Short line method of segment casting

4. Roll new segment into old segment position,
5. Reset forms for next segment and repeat cycle.

Approximately four segments per 5 day week can typically be produced. The main characterising feature of the method is that the required camber is obtained by positioning the old segment relative to the die to be cast. Thus a strong rigid unit rather than the relatively unstable formwork is adjusted. Also the soffit of the new segment can occupy a permanent geometrical position.

The setting out positions are usually determined by calculations from the theoretical casting curve, but graphical presentation offers a more convenient check, especially for comparing as-cast positions against the theoretical.

Long Line Method

Occasionally where space permits the segments are cast as a bed equal in length to the cantilevering arm and laid out to give the geometry of the theoretical casting curve (fig. 36.58). Rail-mounted formwork is required together with its associated curing equipment. The foundation must also be firmly secured to avoid settlement during concreting operations. Segments having obtained sufficient strength can be removed to a stockyard, freeing the bed for production of the next half span.

DIRECTION OF MOBILE
FORMWORK SYSTEM

Figure 36.58 Long line casting method

PREFABRICATED STEEL OR CONCRETE SEGMENTS STAYED BY CABLES

Cable-stayed bridges are proving popular for spans up to about 800 m and when coupled with approach roads very broad crossings are bridgeable. Precast or *in situ* concrete or steel box segments may be used for the deck, which is supported by cables and stayed to a tower. There are four principle configurations, known as radiating (fig. 36.62a), harp (fig. 36.62d), fan and star systems. These arrangements provide compression in the deck by utilising self-weight as shown in fig. 36.59. In this way the deck can be made up of individual segments and made to act like a prestressed beam.

Clearly cables can be spaced such that the horizontal component of force cancels out any tensile force in either the top or bottom flange, individual segments can then be assembled one-by-one and left unconnected. In the case of a concrete deck, shear forces would be resisted by shear keys and/or epoxy resin glued joints, whereas steel would normally be welded.

Where the cables are much further apart, then the tensile stresses would have to be removed by prestressing or carried in the flange of a continuous steel box girder. Vertical force components in each cable are transferred to the pylon and carried through to a foundation.

367

Figure 36.59 Principles of the cable-stayed technique

STABILITY CONDITIONS

The principle aim of the structural configuration of a cable-stayed arrangement, is to prevent sideways and vertical movements of the tower/pylon and deck under asymmetrical live loading. By careful selection of the foundation types and connection of cable and girder it is possible to maintain stability of the whole structure by resisting only the horizontal and vertical components of the forces generated, for example the case of instability illustrated in fig. 36.60 with a live load W placed at point A causing rotation of the girder about the top of the pylon, is countered as illustrated in fig. 36.61 (i). Here the girder resists compressive forces, the cables are pin-pointed to the deck and tower, the tower and its foundation are hinged to the deck and the end supports are free to move horizontally but not vertically. Thus the same load at A (fig. 36.61 ii) now resolves into equilibrium of the first order and the whole system is stable.

Figure 36.60 An unstable arrangement

Figure 36.61 A stable
arrangement

Figure 36.62 Examples
of stable and unstable
configurations

Other stable and non-stable arrangements are shown in figs. 36.62 (a) to (f) including (c) which has a fixed earth support, horizontal movement of the girder being resisted by abutments.

Not all systems, however, are designed as stable of the first order, for example the harp arrangement (d) and (e). Here the partial system ABC is inherently unstable unless either the deck or pylon are capable of resisting bending or alternatively supports placed under each stay for one-half deck span (f). In practice, designers often choose a mixture of stiff deck and pylon, coupled with a range of supports from the self-anchoring type to fully fixed.

TRANSVERSE ARRANGEMENT OF CABLES

Viewed perpendicularly to the line of the bridge, the cables are usually either arranged in a single-plane or two-plane system (fig. 36.63). Single-plane is commonly employed with a divided road deck, and requires only a narrow pylon and pier. The deck itself generally has a hollow box cross section to provide torsional resistance across the deck width. In the two-plane system the cable can either be arranged to hand vertically or slope towards the top of the tower or pylon, the connection to the deck being through the outside edges.

Figure 36.63 Typical transverse cable arrangements

CABLES

SINGLE PLANE   TWO-PLANE

THE PYLON (TOWER)

The pylon may be fabricated from steel plate, or precast concrete elements or occasionally in *in situ* concrete. The various configurations shown, in fig. 36.64 illustrate the flexibility of design options available to produce good aesthetic effect.

THE DECK

Like the pylon, the superstructure may be assembled in precast concrete elements, steel plate or girders, or made in *in situ* concrete. The most common form being the box section, which offers good torsional restraint (fig. 36.65). Plate girders (fig. 36.66) are sometimes used with a double plane system of hangers, where erection procedures require assembly in small light elements. Trusses are also an option but the high fabrication costs, expensive mainten-ance to conteract corrosion and poor aerodynamic characteristics now render this method relatively uneconomic.

Figure 36.64 Typical tower configurations

SINGLE TOWER

TWIN TOWER

A-FRAME TOWER

DIAMOND TOWER

WEB CABLE

WELDED PLATE

STIFFENING MEMBER CABLES

CABLE

CONCRETE BOX FRAME

STIFFENING MEMBER

WELDED PLATE

Figure 36.65 Typical box section deck arrangement in steel plate

STAY CABLES

Figure 36.66 Girder deck

Figure 36.67 Parallel wire mono-strand cable

THE CABLE AND CONNECTIONS

The cable material is similar to that used for normal prestressing work and either comprises multi-strand cable (fig. 30.24) made up of cold drawn wires as described in chapter 30 or alternatively as single strand cable (mono-strand cable) consisting of parallel wires (fig. 36.67). Diameters in the range 40–125 mm are typical. Protection against corrosion can be provided by galvanising each wire, but a more thorough practice has been to cover the cable in steel or plastic ducting and subsequently inject cement grout after positioning in place (fig. 36.68). This latter operation is carried out after all dead loads have been applied to avoid too much·cracking of the mortar.

Figure 36.68 Protection and sheathing of mono-strand cable

Figure 36.69 Pin jointing of mono-strand cable to tower

The cable is normally connected to the pylon with pin-type joints as illustrated in the examples shown in fig. 36.69 or alternatively placed in the groove or guide tube of a saddle (fig. 36.70), depending upon the design requirements.

The cable ends for the pin-type connection have either swaged (fig. 36.71) or filled sockets. Swaging consists of squeezing a socket onto the wire in a hydraulic press and is generally used with strand having a diameter in the range 10–40 mm. Filled sockets are more suited to the larger diameter parallel wire type cable with the socket containing the whole bundle of wires. Several alternative types are manufactured differing slightly in the form of dead ending of each wire and the type of filling material. In the most simple form the wires are led through a plate at the base of the socket and finished with a button head or sockets and wedge. The inside of the socket, conical

Figure 36.70 Saddle for cable on tower

Figure 36.71 Swaged socket for pin-jointed cable

in shape, is subsequently filled with an alloy of zinc, copper, alluminium or lead, or sometimes with a cold casting compound such as epoxy resin. Thus when the cable is subject to a tension load, wedging action develops thereby increasing the grip on the wires. The deck-to-cable connection is usually of the 'free' type to accommodate adjustment. A flared arrangement is required for multi-strand cable (fig. 36.72), while only a single socket is usually needed for mono-strand cable (fig. 36.73). Initial tensioning of the cable to remove slack is generally carried out with a hydraulic jack similar to that used in prestressed concrete, the socket is therefore often manufactured with an internal thread for the jack connection and external thread and nut to take up the extension and other adjustments. Figure 36.74 illustrates a typical example, but other types allowing a pin joint are manufactured.

Figure 36.72 Flared joint for multi-strand cable connection to deck

Figure 36.73 Connection of mono-strand cable to deck

Figure 36.74 Mono-strand cable/deck connection detail

METHODS OF ERECTION

The appropriate method of erection is influenced by the stiffness of the pylon, cable anchorage system, viability of installing temporary supports, maximum unsupported spans permitted by the design, ease of transporting materials, etc. However, since stability of the system largely depends upon transferring the horizontal component of the force in a cable through the stiffening girder, it is clearly necessary to have girder continuity between each pair of stays.

The different procedures commonly adopted to ensures this are:

1. Erect on temporary props.
2. Free cantilever with progressive placing.
3. Balanced cantilever.
4. Push-out.

Erect on Temporary Props (Staging Method)

This method is appropriate when the pylon is not designed with full end fixity to the pier or cannot be temporarily fixed, i.e. the pylon is not stable unless the anchor cable is held in position. Figure 36.75 illustrates a typical erection procedure beginning at one of the abutments. Temporary piers are first installed and the deck units progressively placed one-by-one and welded together to form short free cantilevers. A derrick-type crane mounted on rail track is commonly used for lifting and thus the weight of a unit would normally have to be significantly less than the derrick capacity (typically about 150 tonnes at minimum radius), and it may sometimes even be necessary for assembly to be carried out in sections. Prefabrication normally takes place off site, and units are erected in 5–15 m lengthss. The length of free cantilever possible during the construction phase depends on the deck characteristics and must be carefully determined for the temporary conditions

Figure 36.75 Staging method using temporary supports

but over 50 m of unpropped section have been successfully achieved. A similar procedure using precast concrete could be used but because of the much heavier weights involved, either shorter sections or specialised lifting carriages would be necessary until the stays were in position.

On completion of the deck, all the stays are connected, tensioned and the temporary piers dismantled. However, some extension of the cable is unavoidable as the self-weight of the deck is taken up. The temporary propping should therefore be erected at a height calculated to allow for this movement.

Free Cantilever with Progressive Placing

In many situations the installation of temporary supports would be difficult and expensive and cantilever construction might be considered as an alternative. Figures 36.76 and 36.77 show a typical example whereby the side spans are constructed on temporary propping followed by the tower. This part of

Figure 36.76 Free cantilever method for a stayed deck using temporary supports

Figure 36.77 Free cantilever sequences with stayed steel or precast concrete segments

STAGE 1

CRAWLER CRANE

STAGE 2

STAGE 3

STAGE 4

STAGE 5

STAGE 6

FINAL POSITION OF
TEMPORARY CABLE
STAYS

STAGE 7

ALL BOXES AND CABLES ERECTED

STAGE 8

the bridge is often situated on the embankments where access may favour the use of cranes at ground level.

The centre span is thereafter erected unit-by-unit working out as a free cantilever from the tower or pylon. Like in the previous method, steel box sections up to 20 m long are commonly lifted either by derrick (fig. 36.78) or with mobile lifting beams (fig. 36.51) and welded into place. Thereafter the permanent stays are fixed each side of the tower and the bending moment caused by the cantilevering section removed. Figure 36.79 demonstrates the use of temporary stays to prevent overstressing of the previously positioned permanent stay and the girder.

The provision of temporary stays is particularly important with precast concrete segments where units weighing up to 300 tonnes are occasionally

Figure 36.78 Typical erection equipment with free cantilever method

Figure 36.79 Precast concrete segment erection using temporary stays

erected. The normal procedure is to match cast adjacent segments and subsequently glue the joints with epoxy resin, temporary post-tensioning being applied to bring the two elements together (fig. 36.80). The permanent cable is tensioned simultaneously as the temporary stay is released.

An *in situ* concrete cable-stayed deck constructed with a mobile carriage and formwork similar to that used in cantilever construction (fig. 36.81) is an alternative to steel and precast concrete, but a rate of progress of one 3–4 m section each week is very slow and thus is more commonly adopted as an alternative to stepping formwork systems on multispan bridges in the range 30–70 m between piers (fig. 36.82).

Figure 36.80 Precast concrete segment erection using temporary prestressing

Figure 36.81 Mobile carriage for *in situ* concrete construction using temporary stays

Figure 36.82 Mobile carriage construction with mobile stay arrangement

Figure 36.83 Precast segment erection with mobile stay arrangement

The cable-staying technique using temporary stays only has also proved successful for multi short span bridges of the precast type (fig. 36.83). This progressive erection method allows units to be transported along the previously constructed deck, which are then swung round and attached to lifting equipment such as swivel arm. The stays are usually tensioned with built-in hydraulic jacks, and the whole device moved forward from pier to pier as each span is erected and post-tensioned.

Balanced Cantilever

The occasional need to have clear uninterrupted space below the bridge, for example railway sidings, private property, etc., has forced designers and constructors to develop the balanced cantilevering technique, whereby all or at least very few props are required, as shown in fig. 36.84. Erection proceeds simultaneously each side to the tower, with the first few sections over the piers, temporarily supported on falsework until the tower has been erected and the cables attached. Like the other methods, a degree of cantilevering beyond the last attached cable may be possible depending upon the capability of the section to resist bending movement, the potential for this possibility being much better for steel plate than heavy precast concrete segments.

Figure 36.84
Balanced cantilever segment erection sequences for stayed deck

An important feature of this technique is the need to have a stiff tower and fixity between the deck and tower and its foundations, because of imbalances caused by construction plant, variation in segment dead weight, and tension in the cables. Where possible, the tower design should be selected to accommodate this requirement, otherwise substantial extra staying, temporary anchor cables or a heavy deck tower fixing clamp must be provided. Cantilever spans over 150 m each side of the tower are commonly erected, but where ever possible some propping is desirable to aid stability.

Push-out Method

In some situations access beyond the abutment may not be available or deck units cannot be transported to the tower over adjoining property. To overcome these difficulties a few bridges have used the push-out method as illustrated in fig. 36.85. The deck is assembled at one of the abutments and simply winched out over the rollers or teflon pad bearings.

A similar technique has been used with incremental launching when temporary cable stays are used rather than props (compare with fig. 36.86).

BEARINGS

Figure 36.85 Push-out method using permanent stays

TEMPORARY TOWER AND STAYS

STEEL NOSE

ROLLER BEARINGS

Figure 36.86 Incremental launching method assisted with temporary stays

CABLE ERECTION

The majority of cable-stayed bridges are nowadays designed with monostrand cable either of the parallel wire or locked coil wire type. A complete stay is manufactured in its polyethelene tubing and delivered to site on reels. The simplest erecting procedure is to unreel the cable along the deck and hoist or lift it up to the top of the tower. Unfortunately the natural sag tends to be quite large and therefore considerable take-up has to be provided in the tensioning jack. A more satisfactory procedure is to install a guide rope and pull the cable up with a hauling rope. Intermediate supports to reduce sag are provided by intermittently spaced sliding hangers (fig. 36.87). Tensioning

Figure 36.87 Stay
cable erection and
positioning

is initially carried out at the deck connection end to take up the stack, final tensioning to remove bending moment in the deck and transfer dead load into the cable being supplied after all work on the newly erected section is complete (i.e. welding, post-tensioning of concrete segments, etc.). The jacking equipment is similar to that used for prestressed concrete, using the threaded bar system (fig. 30.19). Finally the ducting is filled with pumped grout to provide protection against corrosion.

FABRICATION OF STEEL DECK UNITS

Steel box sections are ideally suited to modern fabrication methods. In particular automatic numerically controlled cutting, drilling, milling and welding machines are positive encouragements towards manufacturing as much of the deck as possible under workshop conditions and bringing finished units to the site. Furthermore recent advances in welding technology such as submerged arc, CO_2, etc., (described in chapter 34) make it possible to perform high quality welds quickly in the field thereby facilitating assembly in manageable size components without loss of performance and quality.

The time required to erect and weld deck units into place depends upon the amount, type of weld, plate thickness, etc., but a 15 m length section can be typically installed in a 2 week (10 day) period.

SUSPENSION BRIDGES

Suspension bridges are suited to spans exceeding 500 m, and comprise cables slung over two towers. The deck units are hung from the cables and connected together to form a stiffening girder. The cable ends are usually earth anchored while the girder generally has fixed hinges at the piers and movable ones at the towers (fig. 36.88). Thus a live load on the centre span will cause deflection of the deck and pylons and a small change in the position of the suspension cable.

THE TOWER

Depending upon the height and mode of erection, towers may be shop-fabricated in steel as complete units or made up from cellular or box girder sections. Occasionally *in situ* concrete either cast lift-by-lift or slip-formed offers an alternative method.

Figure 36.88 Principle of the suspension bridge arrangement

CABLE

Except on very short spans the suspension cable is generally spun in place from individual galvanised wires, or alternatively positioned similar to the method used for cable-stayed bridges. Both systems require the wire or stands to be compacted together and then bound in galvanised wire and coated with weather-resistant paint to aid corrosion protection.

HANGERS

Hanger cables are clamped directly to the suspension cable with tightly bolted bands. The main cable wrapping only extends up to the bands as these are fixed prior to the wrapping operation.

DECK UNIT

Two types of deck system are used for suspension bridges, namely:

1. Trussed girders.
2. Steel box sections.

Trussed girders were predominantly used on the earlier bridges with the members erected piece-by-piece from a crane running on the previously erected deck. Alternatively the trusses were assembled into large sections and lifted into place. The high costs of fabrication, inefficient use of steel inherent in the design, the large quantities of labour-intensive work and maintenance costs, however, forced designers towards cellular steel box sections (fig. 36.89). These can be prefabricated cheaply and quickly delivered to site, complete with footpaths and railing. They are easily connected together with bolts and/or welding.

Figure 36.89 Typical steel box section deck

Once the units are joined, the concrete surfacing is laid and the deck finishes completed.

ANCHORAGES

The anchorage is fundamental to the stability of a suspension bridge, as all the load in the cable must be transferred to a fixed anchorage (a few small bridges with self-anchorages similar to cable-stayed bridges have been used). There are commonly three types:

1. Rock anchors.
2. Tunnel anchorage.
3. Gravity anchorage.

Rock anchorages simply involve drilling into the rock and grouting in large bolt-type anchors to which the strands are subsequently attached. Where suitable rock is available, a U-tunnel can be constructed and the two cables joined to form a loop (fig. 36.90). However, the gravity anchorage has proved to be most popular with designers. The basic arrangement shown in fig. 36.91 consists of looped over strand shoes attached to anchor bolts located in the concrete. The cable forces are thus resisted by a combination of overburden, dead weight and bearing friction.

Figure 36.90 Tunnel anchorage

Figure 36.91 Gravity anchorage

SADDLES

The cable is usually spun in place, strand-by-strand. When the required number of wires in a strand is achieved they are grouped together with straps or bands placed at about 2 m intervals. The saddle at the top of each tower and the splay saddles are cast in steel and grooved, stepped and divided with

Figure 36.92 Saddles at the top of the towers

COVER PLATE

STEEL CASTING

SPACER PLATES

GROOVES

STEEL PLATE

ELEVATION

SECTION A-A

PLAN

steel spacers to aid wire location during spinning (fig. 36.92). A cover plate is provided for protection against corrosion. The whole unit is bolted down to resist movement.

ERECTION METHODS

The two systems most commonly selected for the deck placing sequences are:

1. Start erection from centre of of main span (fig. 36.93).
2. Start erection from the towers (fig. 36.94).

The deck section joints are left unconnected until the last unit is in position as significant displacement of the main cable and towers takes place while the dead loading is progressively increased. Calculations should therefore be

LIFTING BEAMS

SAGGING

CABLE CURVATURE APPROACHING DESIGN SHAPE

Figure 36.93 Erection from centre of main span method

Figure 36.94 Erection from towers method

made of the cable tension and shape, and distortion of the towers during the erection sequence to ensure tolerable limits of deformation are maintained. A typical sequence of operations is as follows:

(i)   Construct piers and anchorages.
(ii) Erect towers and saddles.
(iii) Hoist catwalk cables into place and build catwalk.
(iv) Place cross bridges and fix storm ties to stabilise catwalk.
(v)   Haul out tramway ropes for spinning wheels.
(vi) Erect suspension cable by air spinning method adjusting wires and strands for sag within tolerance specification.
(vii) Compact cable and temporarily clamp.
(viii) Fix permanent bands and hangers to approximate positions operating from work carriages running on the tramway cable.
(ix) Erect deck units.
(x)   Connect deck joints when all units are in place and complete tightening of hanger bands.
(xi) Wrap main cable.
(xii) Complete road surfacing, etc.

TOWER ERECTION

The bridge towers can be assembled in segments or formed from *in situ* concrete, using either the slip-form method or climbing forms. A tower crane (fig. 36.95) or derrick on gabbards is normally placed at each leg for the purpose.

Figure 36.95 Tower erection with climbing cranes

The crane can also assist in placing the saddles and hoisting up the free ends of the catwalk cable strands. A passenger/materials hoist clipped to the side of the tower is also generally provided for rapid transfer of men and materials.

CATWALK AND HAULING SYSTEM

Walkways (or catwalks) are provided just below each main cable to give access during the spinning and subsequent erection operation. Typically these are about 3 m wide, consisting of hardwood tread boards mounted on galvanised mesh all carried on wire strands, with two additional strands acting as

Figure 36.96 Catwalk and hauling system

387

handrailing. Stiffening frames are spaced regularly (approximately 50 m) along the span and cross bridges provided to link the two walks. Also two tramway support strands are stretched above the catwalk between the main towers and from towers to abutments to carry the stabilising cross beams (at approximately 90 m spacings) and spinning wheel haul rope sheaves. The whole system is stabilised by means of strong ropes anchored to the towers and abutments. The walkway strands are the first to be placed in position using barges. The free end of each is raised and connected to the top of the tower while the rest is unwound off the reel along the bed of the river, and subsequently fixed to the abutment or other tower. Adjustment is generally necessary to produce the desired sag for the spinning operation.

The rest of the catwalk equipment such as mesh, stiffening frames, main bridging beams, etc., are usually slid down the strands working from the tower tops. A small working carriage running on the handrail strands is generally set up to transport the workmen.

The tramway support strands are next erected in a similar manner to those of the footbridge, with the cross beams also being slid into position under control from the tower tops and tied to the catwalk frames. With these operations complete the tramway haul rope can be winched along the catwalk.

CABLE ERECTION

Reeling Operation

The main cable wire is generally transported to site in coils and wound on to 2 m diameter reels (fig. 36.97) capable of holding about 10–15 span lengths. The end of the wire is led round a tensioning wheel (which turns against a controlled brake) and on to the reel. Ends can be spliced together with swaged couplings.

Figure 36.97 Reeling operation

Cable Spinning

The drums of wire are taken from the reeling area and set up on electrically powered unreeling machines (in the example shown a single reel is illustrated but often up to eight wires are simultaneously unreeled across the span). The end of the wire is passed over a series of pulleys housed in a high tower and led via a floating sheave to a strand shoe, where a temporary connection is made. The wire is then looped by hand over the spinning wheel and hauled with the endless tramway drive rope at 200–300 m/min to a strand shoe on the anchorage at the other bank (fig. 36.99). Thus two wires are simultaneously spun, one dead (i.e. that fixed to the shoe) and the other live. The dead wire is temporarily held in position with hooks spaced at approximately 150 m

Figure 36.98 Cable spinning method

Figure 36.99 Cable anchorage bolt and strand shoe

centres along the catwalk, while the live wire is run through sheaves to aid control. Meanwhile the empty wheel returns to the reeling side. Adjustment of sag for each wire generally takes place while spinning is in operation. As soon as the wheel passes over a tower the dead wire on the side span is unhooked, pulled clear of the catwalk with an electric winch, lined up with a guide wire and clamped at the top of the tower, it is then manually lifted into the strand formers located along the catwalk at about 150 m centres. The procedure is then repeated in the main span after the wheel has passed the other tower. Adjustment of the remaining side span sag takes place after the wire has been looped over the strand shoe while the wheel is stationary in ready uses for the return journey to the reeling side. During this return period, the previously live wire is released from the sheaves and adjusted in a similar manner to the dead wire.

The process is repeated until a complete strand has been made, whereafter the loose end is spliced to the temporary connection initially provided for the start end of the wire. The wires are subsequently shaken out, temporarily banded into strands and any final sag adjustments taken up by the strand shoe. The remaining strands are spun in a similar manner until the whole cable is formed. Spinning can usually only take place when wind speeds are less than about 50 km/h. The whole process typically takes up to 6 months for a 1000 m span bridge, with an average about 6 tonnes per day of spun strand.

Cable Compaction
When all the strands are in place, compaction is required to produce the desired shape and reduce the voids. This process is done with a compacting machine, running on the handrail strands and working down from the towers.

The machine encircles the cable and squeezes the wires together with hydraulic jacks, breaking the temporary strand ties in the process. Heavy strapping is then fixed around the cable at 1 m spacing. On completion of these operations the cable is stable and free to hang without the aid of the strand formers.

Cable Banding and Hanger Erection

After removal of the tramway beams, the cable bands (fig. 36.100) for the deck hangers are bolted into the exact position required for each deck unit, a work carriage running on the tramway support strands is generally set up to assist transport. Thereafter the deck hangers are fixed one-by-one, the usual procedure being to unreel the hanger at the top of the tower and allow it to unwind along the catwalk. It is then lowered under winched control and connected to the band. The storm ropes are finally removed and the catwalk slackened off to allow for cable movement during erection of the deck.

Figure 36.100 Cable banding and hanger connection

Deck Erection

The sequence of deck erection is shown in figs. 36.93 and 36.94. Modern practice usually prefers box sections either barged or self-floated into position below the hangers. Great care is required in prefabricating the units to ensure matching joints and prevent distortion during transportation. This can generally be managed with 15–30 m long sections weighing up to 300 tonnes. The box sections are usually lifted at four points using two support beams running on the main cable although occasionally derricks-mounted barges are used. The winches are usually placed on platforms located at the tower or on its foundation.

During the lifting operation the beams are locked to the cable to avoid movement caused by the pull from the winches. Jointing of the units should normally be left until the last section is in place, since hogging or sagging will occur depending upon the erection sequence.

CABLE WRAPPING

The hangers are given a final tightening whereafter the cable is wrapped in galvanised mild steel wire (3–4 mm diameter) to provide corrosion protection. The wrapping machine consists of two drums encircling the cable and the wire is unwound as the unit in winched along. Wrapping can be achieved from either end of the machine in order that wire may be brought tight against a hanger band. The drum should contain just enough wire to cover between two hanger bands and allow the drums to be slit apart to pass the bands. The whole unit is hung from the tramway support strands and travels at about 0.5 m/min. The final operation requires the catwalk and other strands to be dismantled, this is a very difficult and dangerous operation requiring the workmen to raise the mesh panels by hand and winch them to the deck below.

FRAME-TYPE BRIDGES

Until quite recently with the introduction of welded steel box sections and prestressed concrete, bridges spanning more than about 30 m, were generally constructed in either riveted or bolted steel members to form a framed structure. The most common form being the Warren or N-girder type trusses. Indeed the method is still widely adopted in parts of the world where labour is plentiful and not too expensive to employ. Provided that the members are well made such that joints marry easily when brought together with simple drifts (pin used prior to bolting fig 36.101), then relatively unskilled labour can be quickly trained to understand the comparatively simple assembly and erection procedures.

Figure 36.101 Bolts and temporary drifts in spliced connection

TRUSSES

These are commonly pre-assembled on the river bank and transferred to pontoons or a barge-mounted crane along a well constructed jetty. The correct camber can be in-built before bolt tightening by using a template (fig. 36.103). If the truss is assembled piece-by-piece over the river site (fig. 36.104) then control becomes much more difficult, however, a much smaller capacity crane can be used and the transporting phase is less risky, especially in tidal or fast-flowing waters. Approximately one 30 m span can be erected and the deck concreted each month.

ARCH FRAMES

A frame arch can normally be erected by building the bridge out over the water, working from either side towards the centre (fig. 36.105). Where possible, the starting sections are tied back to the abutment or alternatively a foundation provided for temporary cables. The steel members are lifted from barges and placed into position with creeper cranes mounted on bogies running along the top string of the arch. The bridge is then erected panel-by-panel, and line and level precisely controlled by adjusting the ties at each

391

Figure 36.102 Truss erection using barges/pontoons

1. STEELWORK ASSEMBLED ON STILLAGES

KENTLEDGE

2. COMPLETED STEELWORK JACKED ONTO CARRIAGES
   LAUNCHED SO THAT THE NOSE IS OVER WATER
   SUPPORTED ON PONTOON

3. SPAN ERECTED AND TRANSFERRED TO PONTOON
   SPAN FLOATED TO FINAL POSITION AND LOWERED INTO POSITION
   PONTOONS RELEASED

Figure 36.103 Putting camber into trusses

ADJUSTABLE PACKING

CONCRETE PADS

Figure 36.104 Erecting trusess piece-by-piece over water

CANTILEVER | CANTILEVER | DERRICK ON GABBARD | CANTILEVER | CANTILEVER | CANTILEVER

HWL

CRANE, LIFTING BEAM etc.

TEMPORARY ERECTION CABLES

CANTILEVERING OUT FROM SHORE
PIN JOINT

TEMPORARY ERECTION CABLES

Figure 36.105 Arch frame construction

COMPLETION OF ARCH

ABUTMENT

abutment. Care is required to ensure that measurements are taken at similar temperatures, wind forces, etc. On meeting at the centre the two halves are finally joined by jacking out the temporary cables at each abutment. Today this bridge type is rarely considered, because cable-stayed or suspensed cable design are usually more economic for the large spans involved.

BIBLIOGRAPHY: SECTION 4

1. Cary, H. B. (1979) *Modern Welding Technology*. Prentice-Hall.
2. Davies, A. C. (1977) *The Science and Practice of Welding*. Cambridge University Press.
3. *Handbook of Operating Instructions for Gas Welding and Cutting* (1984) London: BOC.
4. MacGillivray, G. I. (1982) *The Kessock Experience*. Students Papers Competition, ICE.
5. Romans, D. and Simons, E. N. (1968) *Welding processes and Technology*. Pitman.
6. Chris, S. C. (1977) Bibliography data on cable-stayed bridges. *Journal of the Structural Division ASCE*, Vol. 103, No. 10.
7. *Concrete Bridge Design* (1969) American Concrete Institute. Publication SP-23.
8. Degenkolb, O. H. (1977) *Concrete Box Girder Bridges*. American Concrete Institute.
9. Gimsin, N. J. (1983) *Cable Supported Bridges* (with a comprehensive bibliography). Wiley.
10. Heins, C. P. and Firmage, D. A. (1978) *Design of Modern Steel Highway Bridges*. Wiley.
11. Lampert, P. (1977) Selection of concrete bridge systems. *Canadian Journal of Civil Engineering*, No. 4.
12. Mathivat, J. (1983) *The Cantilever Construction of Prestressed Concrete Bridges*. Wiley.
13. Podolny, W. and Scalzi, J. B. (1951) *Modern Bridge Construction* (2nd edn). Technical Press.
14. *The Forth Bridge and its Builders*, (1967) Institution of Civil Engineers.
15. Troitsky, M. S. (1977) *Cable-stayed Bridges – Theory and Design*. Crosby Lockwood Staples.
16. Wittfoht, H. (1980) *Bridges – Construction Techniques and Methods (Concrete and Steel)*. IABSE Proceedings.
17. Sir Alexander Gibb and Partners. *Various literature*. Reading.
18. RDL International. *Various literature*. Bedford.

INDEX